Number Theory

Number Theory

Jude Randall

Larsen & Keller
www.larsen-keller.com

Number Theory
Jude Randall
ISBN: 978-1-64172-415-9 (Hardback)

Larsen & Keller
Published by Larsen and Keller Education,
5 Penn Plaza,
19th Floor,
New York, NY 10001, USA

Cataloging-in-Publication Data

Number theory / Jude Randall.
p. cm.
Includes bibliographical references and index.
ISBN 978-1-64172-415-9
1. Number theory. 2. Algebra. 3. Mathematics. I. Randall, Jude.
QA241 .N86 2020
512.7--dc23

Printed and bound in China.

For more information regarding Larsen and Keller Education and its products, please visit the publisher's website www.larsen-keller.com

TABLE OF CONTENTS

PREFACE

This book has been written, keeping in view that students want more practical information. Thus, my aim has been to make it as comprehensive as possible for the readers. I would like to extend my thanks to my family and co-workers for their knowledge, support and encouragement all along.

The branch of pure mathematics that primarily deals with the study of the integers and integer-valued functions is referred to as number theory. It includes the study of prime numbers as well as the properties of objects made out of integers. There can be various sub-divisions of number theory. The main sub-divisions of number theory are analytical number theory, algebraic number theory and Diophantine geometry. Some examples of the problems in analytical number theory are the prime number theorem, the Warring problem, Goldbach conjecture, and the Riemann hypothesis. Some of the other subfields of number theory are probabilistic number theory, arithmetic combinatorics and computational number theory. This textbook contains some path-breaking studies on number theory. It traces the progress of this field and highlights some of its key concepts and applications. This book will serve as a reference to a broad spectrum of readers.

A brief description of the chapters is provided below for further understanding:

Chapter – Introduction

The branch of mathematics that deals with the study of the integers and integer-valued functions is called number theory. Elementary number theory refers to the branch of number theory that uses arithmetic algebra and geometry for solving integer and rational solutions. This is an introductory chapter which will briefly introduce about number theory.

Chapter – Arithmetic Functions

Arithmetic functions are real and complex valued functions that are defined on a positive set of integers. It includes additive function, multiplicative function, divisor function, Bell series, Lambert series, Dirichlet convolution, Liouville function, etc. This chapter closely examines these arithmetic functions to provide an easy understanding of the subject.

Chapter – Analytic Number Theory

Analytic number theory uses various methods from mathematical analysis for solving problems related to integers. Prime number theorem, totient function, prime triplet, prime quadruplet, Waring's problem, Sierpinski number, etc. are some of its concepts. This chapter has been carefully written to provide an extensive understanding of the subject.

Chapter – Algebraic Number Theory

The concepts that fall under the domain of algebraic number of algebraic number theory include Gaussian integer, Cubic field, biquadratic field, roots of unity, class field theory, ideal class group, etc. The topics elaborated in this chapter will help in providing a thorough understanding of these concepts of algebraic number theory.

Chapter – Combinatorial Number Theory

Combinatorial number theory is the combination of combinatorics and number theory. It includes restricted sumset, Gilbreath's conjecture, Znám's problem, Szemerédi's theorem, etc. This chapter closely examines about these concepts related to combinatorial number theory to provide an easy understanding of the subject.

Jude Randall

Introduction

CHAPTER 1

The branch of mathematics that deals with the study of the integers and integer-valued functions is called number theory. Elementary number theory refers to the branch of number theory that uses arithmetic algebra and geometry for solving integer and rational solutions. This is an introductory chapter which will briefly introduce about number theory.

Number theory is a branch of mathematics concerned with properties of the positive integers (1, 2, 3, ...). Sometimes called "higher arithmetic," it is among the oldest and most natural of mathematical pursuits.

Number theory has always fascinated amateurs as well as professional mathematicians. In contrast to other branches of mathematics, many of the problems and theorems of number theory can be understood by laypersons, although solutions to the problems and proofs of the theorems often require a sophisticated mathematical background.

Until the mid-20th century, number theory was considered the purest branch of mathematics, with no direct applications to the real world. The advent of digital computers and digital communications revealed that number theory could provide unexpected answers to real-world problems. At the same time, improvements in computer technology enabled number theorists to make remarkable advances in factoring large numbers, determining primes, testing conjectures, and solving numerical problems once considered out of reach.

Modern number theory is a broad subject that is classified into subheadings such as elementary number theory, algebraic number theory, analytic number theory, geometric number theory, and probabilistic number theory. These categories reflect the methods used to address problems concerning the integers.

ELEMENTARY NUMBER THEORY

Elementary number theory is the branch of number theory that investigates properties of the integers by elementary methods. These methods include the use of divisibility properties, various forms of the axiom of induction and combinatorial arguments. Sometimes the notion of elementary methods is extended by bringing in the simplest elements of mathematical analysis. Traditionally, proofs are deemed to be non-elementary if they involve complex numbers.

Usually, one refers to elementary number theory the problems that arise in branches of number theory such as the theory of divisibility, of congruences, of arithmetic functions, of indefinite equations, of partitions, of additive representations, of the approximation by rational numbers, and of continued fractions. Quite often, the solution of such problems leads to the need to go beyond the framework of elementary methods. Occasionally, following the discovery of a non-elementary solution of some problem, one also finds an elementary solution of it.

As a rule, the problems of elementary number theory have a history going back over centuries, and they are quite often a source of modern trends in number theory and algebra.

From preserved ancient Babylonian cuneiform tablets one may deduce that the Babyloneans were familiar with the task of factoring natural numbers into prime factors. In the 5th century B.C. the Pythagoreans established the so-called doctrine of even and odd numbers and justified the proposition that the product of two natural numbers is even if and only if at least one of the factors is even. A general theory of divisibility was created, in essence, by Euclid. In his Elements (3rd century B.C.), he introduces an algorithm for finding the greatest common divisor of two integers and on this basis he justifies the main theorem of the arithmetic of integers: Every natural number can be factored in one and only one way into a product of prime factors.

After C.F. Gauss, at the beginning of the 19th century, constructed a theory of divisibility of complex integers, it became clear that the study of an arbitrary ring must begin with the construction of a divisibility theory in it.

All properties of the integers are connected in one way or another with the prime numbers. Therefore, questions on the disposition of the prime numbers in the sequence of natural numbers evoked the interest of scholars. The first proof that the set of prime numbers is infinite is due to Euclid. Only in the middle of the 19th century did P.L. Chebyshev take the following step in the study of the function $\pi(x)$, the number of prime numbers not exceeding x. He succeeded in proving by elementary means inequalities that imply,

$$0.92129\frac{x}{\operatorname{In}x} < \pi(x) < 1.10555\ldots\frac{x}{\operatorname{In}x}$$

for all sufficiently large x. Actually, $\pi(x) \sim x / \operatorname{In}x$, but this was not established until the end of the 19th century by means of complex analysis. For a long time it was considered impossible to obtain the result by elementary means. However, in 1949, A. Selberg obtained an elementary proof of this theorem. Chebyshev also proved that for $n \geq 2$ the interval $(n,2n)$ contains at least one prime number. A refinement of the interval containing at least one prime number requires a deeper study of the behaviour of the function $\pi(x)$.

In the 3rd century B.C. the sieve of Eratosthenes was used to select the prime numbers from the set of natural numbers. In 1918 V. Brun showed that a modification of this method can serve as a basis for the study of a set of “almost primes”. He proved the Brun theorem on prime twins. The Brun sieve is a special case of general sieve methods which give estimates for collections of almost primes not exceeding x and belonging to a sequence $\{a_n\}$ of natural numbers. Sieve methods can be used when “good” approximations in the mean are known for the amount of integers $a_n \leq x$ belonging to a progression the modulus of which grows with x. Among the sieve methods developed after Brun, the Selberg sieve is of special significance. The strongest results are obtained by a combination of sieve methods and analytic methods. A sieve method in conjunction with the

Shnirel'man method made it possible to effectively find a number k such that any natural number $n \geq 4$ can be represented as a sum of at most k prime numbers.

Two integers a and b are said to be congruent modulo $m \geq 1$ if they have the same remainder on division by m. Gauss introduced the notation $a \equiv b \pmod{m}$. This form of writing, which brings about the analogy of congruences and equations, turned out to be convenient and instrumental in the development of the theory congruences.

Many results obtained previously by P. Fermat, L. Euler, J.L. Lagrange, and others, and also the Chinese remainder theorem, can be stated and proved simply in the language of the theory of congruences. One of the most interesting results of this theory is the quadratic reciprocity law.

The ancient Babylonians knew a large number of "Pythagorean triples". Apparently they knew some method for finding arbitrary many integer solutions of the indeterminate equation $x^2 + y^2 = z^2$. The Pythagoreans used the formulas $x = (a^2 - 1)/2$, $y = a$, $z = (a^2 + 1)/2$ to find solutions of this equation. Euclid indicated a method that allows one to find in succession integer solutions of the equation $x^2 + 2y^2 = 1$ (a special case of the Pell equation). In his Aritmetika Diophantus (3rd century B.C.) made an attempt to setting up a theory of indeterminate equations. In particular, for the solution of equations of the second and some higher degrees he used systematically a device that enabled him to find from one rational solution of a given equation other rational solutions of it. Fermat (17th century) discovered another method, the "method of descent", and solved by it a number of equations, but the so-called Fermat great theorem, which he declared to have solved, has turned out to be beyond the power of elementary methods.

Fermat solved the problem of representing natural numbers by sums of two squares of integers. As a result of research by Lagrange and Gauss the problem of the representation of integers by a definite binary quadratic form was solved. Gauss developed the general theory of binary quadratic forms. The solution of the problem of representing numbers by forms of higher degree (for example, the Waring problem) and by quadratic forms in several variables usually go beyond the framework of elementary methods. Only certain special cases of such problems can be solved elementarily. An example is Lagrange's theorem: Every natural number is the sum of four squares of integers. It should be mentioned that Diophantus in his Aritmetika repeatedly used the possibility of representing a natural number as the sum of four squares of integers.

To elementary number theory one also reckons the theory of partitions, the foundations of which were laid by Euler. One of the basic problems of the theory of partitions is the study of the function $P(n)$, the number of representations of a natural number n as a sum of natural numbers. Other functions of similar type are also treated in the theory of partitions. Continued fractions appeared in connection with problems of approximate computations (extraction of the square root of a natural number, search for approximations of real numbers by common fractions with small denominators). Continued fractions are applied in solving indefinite equations of the first and second degree. Using the apparatus of continued fractions J. Lambert was the first to establish that the number π is irrational. In the solution of various problems in the approximation of real numbers by rational numbers one uses in elementary number theory the Dirichlet principle, besides continued fractions.

In number theory it is easy to state many problems that can be formulated elementarily and have so far remained unsolved. For example; Is the set of even perfect numbers finite or not? Is there at least one odd perfect number? Is the set of Fermat prime numbers finite or not? Is the set of Mersenne prime numbers finite or not? Is the set of prime numbers of the form n^2+1 finite or not? Is it true that there is at least one prime number between the squares of two consecutive natural numbers? Is the set of incomplete fractions of the expansion of $2^{1/3}$ into a continued fraction bounded or not?

Arithmetic Functions

CHAPTER 2

Arithmetic functions are real and complex valued functions that are defined on a positive set of integers. It includes additive function, multiplicative function, divisor function, Bell series, Lambert series, Dirichlet convolution, Liouville function, etc. This chapter closely examines these arithmetic functions to provide an easy understanding of the subject.

Arithmetic function is any mathematical function defined for integers (..., −3, −2, −1, 0, 1, 2, 3, ...) and dependent upon those properties of the integer itself as a number, in contrast to functions that are defined for other values (real numbers, complex numbers, or even other functions) and that involve various operations from algebra and calculus. Examples of arithmetic functions include the following, which associate with each integer n: (1) the number of divisors of n; (2) the number of ways n can be represented as a sum or product of a specified number of integers; (3) the number of primes (integers not divisible by any number greater than one, except themselves) dividing n (including n itself). Arithmetic functions have applications in number theory, combinatorics, counting, probability theory, and analysis, in which they arise as the coefficients of power series.

Some Notational Conventions

Divisor sums and products: Let $n \in \mathbb{N}$.

- $\sum_{d|n} f(d)$ denotes a sum of $f(d)$, taken over all positive divisors d of n.
- $\sum_{p|n} f(p)$ denotes a sum of $f(p)$, taken over all prime divisors p of n.
- $\sum_{p^\alpha \| n} f(p^\alpha)$ denotes a sum of $f(p^\alpha)$, taken over all prime powers p^α that occur in the standard prime factorization of n. (Here the double bar in $p^\alpha \| n$ indicates that p^α is the exact power of p dividing n, i.e., $p^\alpha | n$, but $p^{\alpha+1} | n$.)
- Products over $d|n, p|n$, etc. are defined analogously.

Empty Sum/Product Convention

A sum over an empty set is defined to be 0; a product over an empty set is defined to be 1. Thus, for example, we have

$$\sum_{p^\alpha \| 1} f(p^\alpha) = 0, \quad \prod_{p^\alpha \| 1} f(p^\alpha) = 1,$$

since there is no prime power p^α satisfying the condition $p^\alpha \| 1$.

The above notational conventions greatly simplify the statements of formulas involving arithmetic functions. For example, using these conventions the rather clumsy formula

$$\varphi(n)=\begin{cases}1 & \text{if } n=1,\\ \prod_{i=1}^{r} p_i^{\alpha_i-1}(p_i-1) & \text{if } n\geq 2 \text{ and } n=\prod_{i=1}^{r} p_i^{\alpha_i} \text{ with distinct primes } p_i \text{ and } \alpha_i \in N,\end{cases}$$

can be rewritten as

$$\varphi(n)=\prod_{p^{\alpha}\|n} p^{\alpha-1}(p-1),$$

without having to introduce subscripts or single out the case $n=1$. (In the latter case, the product is an empty product, so by the empty product convention, it produces the value 1, which is exactly what we need).

Sums Over 1's ("Bateman Summation")

A sum in which each summand is equal to 1 simply counts the number of terms in it; for example, $\sum_{d|n} 1$ is the same as $\#\{d \in N : d|n\}$. While this might seem like a contrived way to represent a counting function, in the context of the general theory of arithmetic functions, such representations are often very useful.

Multiplicative Arithmetic Functions

Definition (Multiplicative arithmetic function): A *function* $f : N \to C$ is called an arithmetic function. An arithmetic function f is called multiplicative if it satisfies the relation,

$$f(n_1 n_2) = f(n_1) f(n_2)$$

whenever $((n_1, n_2)=1)$. If the multiplicative arithmetic function holds for all $n_1, n_2 \in N$ (i.e., without the restriction $(n_1, n_2)=1$), then f is called completely multiplicative.

Proposition (Multiplicative functions and prime factorization): An arithmetic function f that is not identically 0 (i.e., such that $f(n) \neq 0$ for at least one $n \in N$) is multiplicative if and only if it satisfies

$$f(n)=\prod_{p\ \|n} f(p\ )\quad (n \in N).$$

In particular, any multiplicative function f that is not identically 0 is uniquely determined by its values $f(p^{\alpha})$ at prime powers and satisfies $f(1)=1$.

The Euler phi Function and the Carmichael Conjecture

Definition (Euler phi function): The Euler phi function is defined by

$$\varphi(n) = \#\{1 \leq m \leq n : (m,n)=1\}.$$

Proposition (Properties of $\varphi(n)$):

- Multiplicativity: The Euler phi function is multiplicative (though not completely multiplicative).
- Explicit formula: For any $n \in \mathrm{N}$,

$$\varphi(n) = \prod_{p^\alpha \| n} p^{\alpha-1}(p-1) = n\prod_{p|n}\left(1-\frac{1}{p}\right).$$

- (Gauss identity):

$$\sum_{d|n}\varphi(d) = n \ (n \in \mathrm{N}).$$

Conjecture (Carmichael conjecture): Given $n \in \mathrm{N}$, the equation $\varphi(x) = n$ has either no solution $x \in \mathrm{N}$ or more than one solution.

Remark: The Carmichael conjecture has several local (UIUC) connections: Its originator, R.D. Carmichael, spent most of his career as a professor here at the U of I, and the conjecture first appeared as an "exercise" in a textbook on number theory he wrote (and which he presumably assigned to his students). Also, most of the current records on this conjecture are held by Kevin Ford, who earned his PhD here in the mid-1990s and is now back as a professor. In particular, Ford showed that the Carmichael conjecture is true for all $n \leq 10^{1000000000}$. Moreover, for any $k \in \mathrm{N}$ except possibly $k = 1$, there exist infinitely many $n \in \mathrm{N}$ such that the equation $\varphi(x) = n$ has exactly k solutions $x \in \mathrm{N}$. Thus, only the question of whether multiplicity $k = 1$ can occur remains open, and this is precisely the question addressed by the Carmichael conjecture.

The Number-of-divisors Functions

Definition (Number-of-divisors function): The number-of-divisors function is defined by,

$$\upsilon(n) = \#\{d \in \mathrm{N} : d|n\} = \sum_{d|n} 1 \ (n \in \mathrm{N}).$$

This function is often simply called the divisor function; alternate, and more common, notations for it are $d(n)$ (for "divisor") and $\tau(n)$ (for "Teiler", the German word for "divisor").

Proposition (Properties of $\upsilon(n)$):

- Multiplicativity: The function $\upsilon(n)$ is multiplicative (though not completely multiplicative).
- Explicit formula: For any $n \in \mathrm{N}$,

$$\upsilon(n) = \prod_{p^\alpha \| n}(\alpha+1).$$

The Sum-of-divisors Functions and Perfect Numbers

Definition Sum-of-divisors function: The sum-of-divisors function is defined by

$$\sigma(n)=\sum_{d|n} d \quad (n \in \mathrm{N}).$$

Proposition (Properties of $\sigma(n)$):

- Multiplicativity: The function $\sigma(n)$ is multiplicative (though not completely multiplicative).
- Explicit formula: For any $n \in \mathrm{N}$,

$$\sigma(n)=\prod_{p^\alpha \| n} \frac{p^{\alpha+1}-1}{p-1}.$$

Definition (Perfect numbers): An positive integer n is called perfect if it is equal to the sum of its positive divisors $d|n$, with $1 \leq d < n$ (i.e., not counting $d=n$). Equivalently, n is perfect if and only if $\sigma(n)=2n$.

Example: The first 4 perfect numbers are 6(= 1 + 2 + 3), 28(= 1 + 2 + 4 + 7 + 14), 496, and 8128.

Theorem (Characterization of even perfect numbers): An even positive integer n is perfect if and only if it is of the form,

$$n=2^{p-1}(2^p-1),$$

where 2^p-1 is a Mersenne prime.

Corollary: There exist infinitely many even perfect numbers if and only if there exist infinitely many Mersenne primes.

Example: The above four perfect numbers 6, 28, 496, 8128 correspond to the first four Mersenne primes, $2^2 - 1 = 3$, $2^3 - 1 = 7$, $2^5 - 1 = 31$, and $2^7 - 1 = 127$.

The Moebius Function and the Moebius Inversion Formula

Definition (Moebius function): The Moebius function is defined by

$$\mu(n)=\begin{cases} 1 & \text{if } \mathrm{n}=1, \\ (-1)^r & \text{if } \mathrm{n}=\mathrm{p}_1\mathrm{p}_2\ldots\mathrm{p}_\mathrm{r} \text{ with distinct prime } \mathrm{p}_\mathrm{i}, \\ 0 & \text{if } n \text{ is not squarefree, i.e., divisible by a prime } \mathrm{p}^\alpha \text{ with } \alpha>1. \end{cases}$$

Remark: A very similar function is the Liouville function $\lambda(n)$, which is defined as $(-1)^r$, where r is the total number of prime factors of n, with multiple prime factors counted multiple times. If n is squarefree, then $\lambda(n)=\mu(n)$, but if n is not squarefree, then $\mu(n)=0$, while $\lambda(n)=\pm 1$ depending on whether n has an even or an odd number of prime factors.

While the Liouville function has a simpler definition and may seem the more natural of the two functions, the Moebius function is more useful and more important because of results such those below (which would not be valid for the Liouville function).

Proposition (Properties of $\mu(n)$):

- Multiplicativity: The Moebius function is multiplicative (though not completely multiplicative).
- Moebius function identity:

$$\sum_{d|n}\mu(d)=\begin{cases}1 & \text{if } n=1,\\ 0 & \text{otherwise}.\end{cases}$$

Theorem (Moebius inversion formula): If f and g are arithmetic functions satisfying

$$f(n)=\sum_{d|n}g(d) \quad (n\in \mathrm{N}),$$

then,

$$g(n)=\sum_{d|n}f(d)\mu(n/d)=\sum_{d|n}\mu(d)f(n/d) \quad (n\in \mathrm{N}).$$

Algebraic Theory of Arithmetic Function

In this topic we develop an algebraic theory of arithmetic functions based on the notion of a Dirichlet product. This allows us to restate many of the definitions and results encountered earlier in a simple, elegant, and natural form.

We begin by defining some "trivial" arithmetic functions that are needed in this theory and which can be used to build up other arithmetic functions.

Definition (Trivial arithmetic functions): The unit function 1, identity function id, and delta function δ are the arithmetic functions defined as follows:

$$1(n)=1 \quad (n\in \mathrm{N}),$$
$$\mathrm{id}(n)=n \quad (n\in \mathrm{N}),$$

$$\delta(n)=\begin{cases}1 & \text{if } n=1,\\ 0 & \text{otherwise}.\end{cases}$$

All three of these functions are multiplicative (in fact, completely multiplicative).

Definition (Dirichlet product of arithmetic functions): Given two arithmetic functions f and g, the Dirichlet product $f * g$ is the arithmetic function defined by,

$$(f*g)(n)=\sum_{d|n}f(d)g(n/d) \quad (n\in \mathrm{N}).$$

Proposition (Algebraic properties of Dirichlet product): Let f, g, h be arithmetic functions.

- Commutativity: $f*g=g*f$.
- Associativity: $(f*g)*h=f*(g*h)$.
- Identity element: $f*\delta=\delta*f=f$, where δ is defined as above, i.e., $\delta(1)=1$ and $\delta(n)=0$ if $n>1$.
- Dirichlet inverse: If $f(1)\neq 0$, then f has a unique Dirichlet inverse f^{*-1}, in the sense that $f*f^{*-1}=\delta$.
- Preservation of multiplicativity: The Dirichlet product of two multiplicative functions is multiplicative.
- Multiplicativity of inverse: The Dirichlet inverse of a multiplicative function is multiplicative.

Using the Dirichlet product notation, we can now restate many of the definitions, identities, and theorems on arithmetic functions encountered earlier in a simple, elegant, very natural, and easy-to-remember form.

Proposition (Drichlet product versions of identities for arithmetic functions):

- Gauss identity: $\varphi*1=\mathrm{id}$.
- Definition of the divisor function: $\upsilon=1*1$.
- Definition of the sum-of-divisors function: $\sigma=1*\mathrm{id}$.
- Moebius function identity: $\mu*1=\delta$.
- Moebius inversion formula: if $f=g*1$, then $g=f*\mu=\mu*f$.

ADDITIVE FUNCTION

In number theory, an additive function is an arithmetic function $f(n)$ of the positive integer n such that whenever a and b are coprime, the function of the product is the sum of the functions:

$$f(ab)=f(a)+f(b).$$

Completely Additive

An additive function $f(n)$ is said to be completely additive if $f(ab)=f(a)+f(b)$ holds *for all* positive integers a and b, even when they are not coprime. Totally additive is also used in this sense by analogy with totally multiplicative functions. If f is a completely additive function then $f(1) = 0$.

Every completely additive function is additive, but not vice versa.

Examples:

Examples of arithmetic functions which are completely additive are:

- The restriction of the logarithmic function to N.
- The multiplicity of a prime factor p in n, that is the largest exponent m for which p^m divides n.
- $a_0(n)$ - the sum of primes dividing n counting multiplicity, sometimes called sopfr(n), the potency of n or the integer logarithm of n. For example:

  $a_0(4) = 2 + 2 = 4$

  $a_0(20) = a_0(2^2 \cdot 5) = 2 + 2 + 5 = 9$

  $a_0(27) = 3 + 3 + 3 = 9$

  $a_0(144) = a_0(2^4 \cdot 3^2) = a_0(2^4) + a_0(3^2) = 8 + 6 = 14$

  $a_0(2{,}000) = a_0(2^4 \cdot 5^3) = a_0(2^4) + a_0(5^3) = 8 + 15 = 23$

  $a_0(2{,}003) = 2003$

  $a_0(54{,}032{,}858{,}972{,}279) = 1240658$

  $a_0(54{,}032{,}858{,}972{,}302) = 1780417$

  $a_0(20{,}802{,}650{,}704{,}327{,}415) = 1240681$

- The function $\Omega(n)$, defined as the total number of prime factors of n, counting multiple factors multiple times, sometimes called the "Big Omega function". For example:

  $\Omega(1) = 0$, since 1 has no prime factors

  $\Omega(4) = 2$

  $\Omega(16) = \Omega(2 \cdot 2 \cdot 2 \cdot 2) = 4$

  $\Omega(20) = \Omega(2 \cdot 2 \cdot 5) = 3$

  $\Omega(27) = \Omega(3 \cdot 3 \cdot 3) = 3$

  $\Omega(144) = \Omega(2^4 \cdot 3^2) = \Omega(2^4) + \Omega(3^2) = 4 + 2 = 6$

  $\Omega(2{,}000) = \Omega(2^4 \cdot 5^3) = \Omega(2^4) + \Omega(5^3) = 4 + 3 = 7$

  $\Omega(2{,}001) = 3$

  $\Omega(2{,}002) = 4$

  $\Omega(2{,}003) = 1$

$\Omega(54{,}032{,}858{,}972{,}279) = 3$

$\Omega(54{,}032{,}858{,}972{,}302) = 6$

$\Omega(20{,}802{,}650{,}704{,}327{,}415) = 7$

Examples of arithmetic functions which are additive but not completely additive are:

- $\omega(n)$, defined as the total number of different prime factors of n. For example:

  $\omega(4) = 1$

  $\omega(16) = \omega(2^4) = 1$

  $\omega(20) = \omega(2^2 \cdot 5) = 2$

  $\omega(27) = \omega(3^3) = 1$

  $\omega(144) = \omega(2^4 \cdot 3^2) = \omega(2^4) + \omega(3^2) = 1 + 1 = 2$

  $\omega(2{,}000) = \omega(2^4 \cdot 5^3) = \omega(2^4) + \omega(5^3) = 1 + 1 = 2$

  $\omega(2{,}001) = 3$

  $\omega(2{,}002) = 4$

  $\omega(2{,}003) = 1$

  $\omega(54{,}032{,}858{,}972{,}279) = 3$

  $\omega(54{,}032{,}858{,}972{,}302) = 5$

  $\omega(20{,}802{,}650{,}704{,}327{,}415) = 5$

- $a_1(n)$ - the sum of the distinct primes dividing n, sometimes called sopf(n). For example:

  $a_1(1) = 0$

  $a_1(4) = 2$

  $a_1(20) = 2 + 5 = 7$

  $a_1(27) = 3$

  $a_1(144) = a_1(2^4 \cdot 3^2) = a_1(2^4) + a_1(3^2) = 2 + 3 = 5$

  $a_1(2{,}000) = a_1(2^4 \cdot 5^3) = a_1(2^4) + a_1(5^3) = 2 + 5 = 7$

  $a_1(2{,}001) = 55$

  $a_1(2{,}002) = 33$

  $a_1(2{,}003) = 2003$

  $a_1(54{,}032{,}858{,}972{,}279) = 1238665$

$a_1(54{,}032{,}858{,}972{,}302) = 1780410$

$a_1(20{,}802{,}650{,}704{,}327{,}415) = 1238677$

MULTIPLICATIVE FUNCTIONS

From any additive function $f(n)$ it is easy to create a related multiplicative function $g(n)$ i.e. with the property that whenever a and b are coprime we have:

$$g(ab) = g(a) \times g(b).$$

One such example is $g(n) = 2^{f(n)}$.

Summatory functions:

Given an additive function f, let its summatory function be defined by $\mathcal{M}_f(x) := \sum_{n \le x} f(n)$. The average of is given exactly as,

$$\mathcal{M}_f(x) = \sum_{p^\alpha \le x} f(p^\alpha)\left(\left\lfloor \frac{x}{p^\alpha} \right\rfloor - \left\lfloor \frac{x}{p^{\alpha+1}} \right\rfloor\right).$$

The summatory functions over f can be expanded as $\mathcal{M}_f(x) = xE(x) + O(\sqrt{x} \cdot D(x))$ where,

$$E(x) = \sum_{p^\alpha \le x} f(p^\alpha) p^{-\alpha}(1 - p^{-1})$$

$$D^2(x) = \sum_{p^\alpha \le x} |f(p^\alpha)|^2 \; p^{-\alpha}.$$

The average of the function f^2 is also expressed by these functions as,

$$\mathcal{M}_{f^2}(x) = xE^2(x) + O(xD^2(x)).$$

There is always an absolute constant $C_f > 0$ such that for all natural numbers $x \ge 1$,

$$\sum_{n \le x} |f(n) - E(x)|^2 \le C_f \cdot xD^2(x).$$

Let,

$$\nu(x;z) := \frac{1}{x} \# \left\{ n \le x : \frac{f(n) - A(x)}{B(x)} \le z \right\}.$$

Suppose that f is an additive function with $-1 \le f(p^\alpha) = f(p) \le 1$ such that as $x \to \infty$,

$$B(x) = \sum_{p \le x} f^2(p)/p \to \infty.$$

Then $\nu(x;z) \sim G(z)$ where $G(z)$ is the Gaussian distribution function,

$$G(z) = \frac{1}{\sqrt{2\pi}} \int_{-\infty}^{z} e^{-t^2/2}\, dt.$$

Examples of this result related to the prime omega function and the numbers of prime divisors of shifted primes include the following for fixed $z \in \mathbb{R}$ where the relations hold for $x \gg 1$:

$$\#\{n \le x : \omega(n) - \log\log x \le z(\log\log x)^{1/2}\} \sim xG(z),$$
$$\#\{p \le x : \omega(p+1) - \log\log x \le z(\log\log x)^{1/2}\} \sim \pi(x)G(z).$$

Multiplicative Function

In number theory, a multiplicative function is an arithmetic function $f(n)$ of a positive integer n with the property that $f(1) = 1$ and whenever a and b are coprime, then

$$f(ab) = f(a)f(b).$$

An arithmetic function $f(n)$ is said to be completely multiplicative (or totally multiplicative) if $f(1) = 1$ and $f(ab) = f(a)f(b)$ holds *for all* positive integers a and b, even when they are not coprime.

Examples:

Some multiplicative functions are defined to make formulas easier to write:

- $1(n)$: The constant function, defined by $1(n) = 1$ (completely multiplicative).
- $\mathrm{Id}(n)$: Identity function, defined by $\mathrm{Id}(n) = n$ (completely multiplicative).
- $\mathrm{Id}_k(n)$: the power functions, defined by $\mathrm{Id}_k(n) = n^k$ for any complex number k (completely multiplicative). As special cases we have,
    - $\mathrm{Id}_0(n) = 1(n)$,
    - $\mathrm{Id}_1(n) = \mathrm{Id}(n)$.
- $\varepsilon(n)$: the function defined by $\varepsilon(n) = 1$ if $n = 1$ and 0 otherwise, sometimes called multiplication unit for Dirichlet convolution or simply the unit function (completely multiplicative). Sometimes written as u(n), but not to be confused with $\mu(n)$.
- $1_C(n)$, the indicator function of the set $C \subset \mathbb{Z}$, for certain sets C. The indicator function $1_C(n)$ is multiplicative precisely when the set C has the following property for any coprime numbers a and b: the product ab is in C if and only if the numbers a and b are both themselves in C. This is the case if C is the set of squares, cubes, or k-th powers, or if C is the set of square-free numbers.

Other examples of multiplicative functions include many functions of importance in number theory, such as:

- $\gcd(n,k)$: The greatest common divisor of n and k, as a function of n, where k is a fixed integer.

- $\varphi(n)$: Euler's totient function φ, counting the positive integers coprime to (but not bigger than) n.
- $\mu(n)$: The Möbius function, the parity (−1 for odd, +1 for even) of the number of prime factors of square-free numbers; 0 if n is not square-free.
- $\sigma_k(n)$: The divisor function, which is the sum of the k-th powers of all the positive divisors of n (where k may be any complex number). Special cases we have,
  - $\sigma_0(n) = d(n)$ the number of positive divisors of n,
  - $\sigma_1(n) = \sigma(n)$, the sum of all the positive divisors of n.
- $a(n)$: The number of non-isomorphic abelian groups of order n.
- $\lambda(n)$: The Liouville function, $\lambda(n) = (-1)^{\Omega(n)}$ where $\Omega(n)$ is the total number of primes (counted with multiplicity) dividing n. (completely multiplicative).
- $\gamma(n)$, defined by $\gamma(n) = (-1)^{\omega(n)}$, where the additive function $\omega(n)$ is the number of distinct primes dividing n.
- $\tau(n)$: The Ramanujan tau function.
- All Dirichlet characters are completely multiplicative functions. For example,
  - (n/p), the Legendre symbol, considered as a function of n where p is a fixed prime number.

An example of a non-multiplicative function is the arithmetic function $r_2(n)$ - the number of representations of n as a sum of squares of two integers, positive, negative, or zero, where in counting the number of ways, reversal of order is allowed. For example:

$$1 = 1^2 + 0^2 = (-1)^2 + 0^2 = 0^2 + 1^2 = 0^2 + (-1)^2.$$

Therefore, $r_2(1) = 4 \neq 1$. This shows that the function is not multiplicative. However, $r_2(n)/4$ is multiplicative.

Properties

A multiplicative function is completely determined by its values at the powers of prime numbers, a consequence of the fundamental theorem of arithmetic. Thus, if n is a product of powers of distinct primes, say $n = p^a q^b$, then $f(n) = f(p^a) f(q^b)$

This property of multiplicative functions significantly reduces the need for computation, as in the following examples for $n = 144 = 2^4 \cdot 3^2$:

$$d(144) = \sigma_0(144) = \sigma_0(2^4)\sigma_0(3^2) = (1^0 + 2^0 + 4^0 + 8^0 + 16^0)(1^0 + 3^0 + 9^0) = 5 \cdot 3 = 15,$$

$$\sigma(144) = \sigma_1(144) = \sigma_1(2^4)\sigma_1(3^2) = (1^1 + 2^1 + 4^1 + 8^1 + 16^1)(1^1 + 3^1 + 9^1) = 31 \cdot 13 = 403,$$

$$\sigma^*(144) = \sigma^*(2^4)\sigma^*(3^2) = (1^1 + 16^1)(1^1 + 9^1) = 17 \cdot 10 = 170.$$

Similarly, we have:

$$\varphi(144) = \varphi(2^4)\varphi(3^2) = 8 \cdot 6 = 48.$$

In general, if $f(n)$ is a multiplicative function and a, b are any two positive integers, then,

$$f(a) \cdot f(b) = f(\gcd(a,b)) \cdot f(\operatorname{lcm}(a,b)).$$

Every completely multiplicative function is a homomorphism of monoids and is completely determined by its restriction to the prime numbers.

Convolution:

If f and g are two multiplicative functions, one defines a new multiplicative function f * g, the Dirichlet convolution of f and g, by:

$$(f * g)(n) = \sum_{d|n} f(d)g\left(\frac{n}{d}\right).$$

Where the sum extends over all positive divisors d of n. With this operation, the set of all multiplicative functions turns into an abelian group; the identity element is ε. Convolution is commutative, associative, and distributive over addition.

Relations among the multiplicative functions discussed above include:

- $\mu * 1 = \varepsilon$ (the Möbius inversion formula).
- $(\mu\, \mathrm{Id}_k) * \mathrm{Id}_k = \varepsilon$ (generalized Möbius inversion).
- $\varphi * 1 = \mathrm{Id}$.
- $d = 1 * 1$.
- $\sigma = \mathrm{Id} * 1 = \varphi * d$.
- $\sigma_k = \mathrm{Id}_k * 1$.
- $\mathrm{Id} = \varphi * 1 = \sigma * \mu$.
- $\mathrm{Id}_k = \sigma_k * \mu$.

The Dirichlet convolution can be defined for general arithmetic functions, and yields a ring structure, the Dirichlet ring.

The Dirichlet convolution of two multiplicative functions is again multiplicative. A proof of this fact is given by the following expansion for relatively prime $a, b \in \mathbb{Z}^+$:

$$(f * g)(ab) = \sum_{d|ab} f(d)g\left(\frac{ab}{d}\right) = \sum_{d_1|a}\sum_{d_2|b} f(d_1d_2)g\left(\frac{ab}{d_1d_2}\right) = \sum_{d_1|a} f(d_1)g\left(\frac{a}{d_1}\right) \times \sum_{d_2|b} f(d_2)g$$

$$\left(\frac{b}{d_2}\right) = (f * g)(a) \cdot (f * g)(b).$$

Dirichlet series for some multiplicative functions:

- $\sum_{n\geq 1} \frac{\mu(n)}{n^s} = \frac{1}{\zeta(s)},$
- $\sum_{n\geq 1} \frac{\varphi(n)}{n^s} = \frac{\zeta(s-1)}{\zeta(s)},$
- $\sum_{n\geq 1} \frac{d(n)^2}{n^s} = \frac{\zeta(s)^4}{\zeta(2s)},$
- $\sum_{n\geq 1} \frac{2^{\omega(n)}}{n^s} = \frac{\zeta(s)^2}{\zeta(2s)},$

More examples are shown in the article on Dirichlet series.

Multiplicative function over $F_q[X]$:

Let $A = F_q[X]$, the polynomial ring over the finite field with q elements. A is a principal ideal domain and therefore A is a unique factorization domain.

A complex-valued function λ on A is called multiplicative if $\lambda(fg) = \lambda(f)\lambda(g)$ whenever f and g are relatively prime.

Zeta function and Dirichlet series in $F_q[X]$:

Let h be a polynomial arithmetic function (i.e. a function on set of monic polynomials over A). Its corresponding Dirichlet series is defined to be,

$$D_h(s) = \sum_{f \text{ monic}} h(f)|f|^{-s},$$

where for $g \in A$, set $|g| = q^{\deg(g)}$ if $g \neq 0$, and $|g| = 0$ otherwise.

The polynomial zeta function is then

$$\zeta_A(s) = \sum_{f \text{ monic}} |f|^{-s}.$$

Similar to the situation in N, every Dirichlet series of a multiplicative function h has a product representation (Euler product):

$$D_h(s) = \prod_P \left(\sum_{n=0}^{\infty} h(P^n)|P|^{-sn} \right),$$

where the product runs over all monic irreducible polynomials P. For example, the product representation of the zeta function is as for the integers:

$$\zeta_A(s) = \prod_P (1 - |P|^{-s})^{-1}.$$

Unlike the classical zeta function, $\zeta_A(s)$ is a simple rational function:

$$\zeta_A(s) = \sum_f |f|^{-s} = \sum_n \sum_{\deg(f)=n} q^{-sn} = \sum_n (q^{n-sn}) = (1-q^{1-s})^{-1}.$$

In a similar way, If f and g are two polynomial arithmetic functions, one defines f * g, the Dirichlet convolution of f and g, by:

$$(f*g)(m) = \sum_{d|m} f(d)g\left(\frac{m}{d}\right)$$
$$= \sum_{ab=m} f(a)g(b),$$

where the sum is over all monic divisors d of m, or equivalently over all pairs (a, b) of monic polynomials whose product is m. The identity $D_h D_g = D_{h*g}$ still holds.

DIVISOR FUNCTION

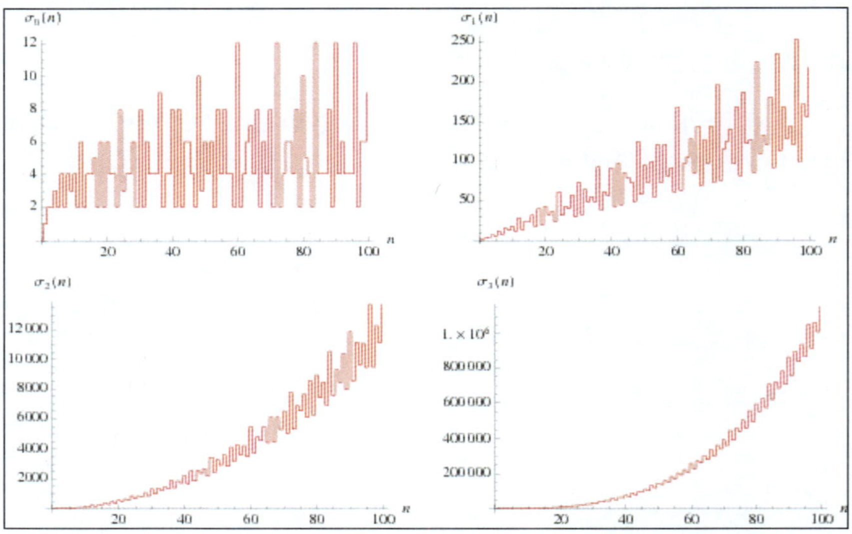

The divisor function $\sigma_k(n)$ for n an integer is defined as the sum of the kth powers of the (positive integer) divisors of n,

$$\sigma_k(n) \equiv \sum_{d|n} d^k.$$

It is implemented in the Wolfram Language as DivisorSigma[k, n].

The notations $d(n)$, $v(n)$, and $\tau(n)$ are sometimes used for ( ), which gives the number of divisors of n. Rather surprisingly, the number of factors of the polynomial $a^n - b^n$ are also given by $d(n)$. The values of $\sigma_0(n)$ can be found as the inverse Möbius transform of 1, 1, 1. Heath-Brown proved that $\sigma_0(n) = \sigma_0(n+1)$ infinitely often. The numbers having the incrementally largest number of divisors are called highly composite numbers. The function $\sigma_0(n)$ satisfies the identities,

$$\sigma_0(p^a) = a+1$$
$$\sigma_0(p_1^{a_1}\, p_2^{a_2} \ldots) = (a_1+1)(a_2+1),$$

where p_i are distinct primes and $p_1^{a_1}\, p_2^{a_2}$ is the prime factorization of a number n.

The divisor function $\sigma_0(n)$ is odd iff n is a square number.

The function $\sigma_1(n)$ that gives the sum of the divisors of n is commonly written without the subscript, i.e., $\sigma(n)$.

As an illustrative example of computing $\sigma_k(n)$, consider the number 140, which have divisors $d_i = 1$ 2, 4, 5, 7, 10, 14, 20, 28, 35, 70, and 140, for a total of $N = 12$ divisors in all. Therefore,

$$\sigma_0(140) = N = 12$$
$$\sigma_1(140) = \sum_{i=1}^{N} d_i = 336$$
$$\sigma_2(140) = \sum_{i=1}^{N} d_i^2 = 27300$$
$$\sigma_3(140) = \sum_{i=1}^{N} d_i^3 = 3164112.$$

The following table summarized the first few values of $\sigma_k(n)$ for small k and $n = 1, 2, \ldots$

k	OEIS	$\sigma_k(n)$ for $n = 1, 2, \ldots$
0	A000005	1, 2, 2, 3, 2, 4, 2, 4, 3, 4, 2, 6, ...
1	A000203	1, 3, 4, 7, 6, 12, 8, 15, 13, 18, ...
2	A001157	1, 5, 10, 21, 26, 50, 50, 85, 91, 130, ...
3	A001158	1, 9, 28, 73, 126, 252, 344, 585, 757, 1134, ...

The sum of the divisors of n excluding n itself (i.e., the proper divisors of n) is called the restricted divisor function and is denoted $s(n)$. The first few values are 0, 1, 1, 3, 1, 6, 1, 7, 4, 8, 1, 16, ...

The sum of divisors $\sigma_1(N)$ can be found as follows. Let $N \equiv ab$ with $a \neq b$ and $(a,b) = 1$. For any divisor d of $N, d = a_i b_i$, where a_i is a divisor of a and b_i is a divisor of b. The divisors of a are 1, $a_1, a_2, \ldots$, and a. The divisors of b are 1, $b_1, b_2, \ldots b$. The sums of the divisors are then,

$$\sigma_1(a) = 1 + a_1 + a_2 + \ldots + a$$
$$\sigma_1(b) = 1 + b_1 + b_2 + \ldots + b.$$

For a given a_i,

$$a_i(1 + b_1 + b_2 + \ldots + b) = a_i \sigma_1(b).$$

Summing over all a_i,

$$(1+a_1+a_2+\ldots+a)\sigma_1(b)=\sigma_1(a)\sigma_1(b),$$

So $\sigma_1 N=\sigma_1(ab)=\sigma_1(a)\sigma_1(b)$. Splitting a and b into prime factors,

$$\sigma_1(N)=\sigma_1\left(p_1^{\alpha_1}\right)\sigma_1\left(p_2^{\alpha_2}\right)\ldots\sigma_1\left(P_r^{\alpha_r}\right).$$

For a prime power $p_i^{\alpha_i}$, the divisors are 1, $p_i, p_i^2, \ldots, p_i^{\alpha_i}$, so,

$$\sigma_1\left(p_i^{\alpha_i}\right)=1+p_i+p_i^2+\ldots+p_i^{\alpha_i}=\frac{p_i^{\alpha_{i+1}}-1}{p_i-1}.$$

For N, therefore,

$$\sigma_1(N)=\prod_{i=1}^{r}\frac{p_i^{\alpha_{i+1}}-1}{p_i-1}.$$

For the special case of N a prime, the above equation simplifies to,

$$\sigma_1(p)=\frac{P^2-1}{p-1}=p+1.$$

Similarly, for N a power of two, equation $\sigma_1(N)=\prod_{i=1}^{r}\frac{p_i^{\alpha_{i+1}}-1}{p_i-1}$ simplifies to,

$$\sigma_1(2^\alpha)=\frac{2^{\alpha+1}-1}{2-1}=2^{\alpha+1}-1.$$

The identities (◇) and (◇) can be generalized to,

$$\sigma_k(N)=\sigma_k\left(p_1^{\alpha_1}\right)\sigma_k\left(p_2^{\alpha_2}\right)\ldots\sigma_k\left(p_r^{\alpha_r}\right)$$
$$=\prod_{i=1}^{r}\frac{p_i^{(\alpha_i+1)\mathrm{k}}-1}{p_i^k-1}.$$

Sums involving the divisor function are given by,

$$\sum_{n=1}^{\infty}\frac{\sigma_0(n)}{n^s}=\zeta^2(s).$$

For $s>1$,

$$\sum_{n=1}^{\infty}\frac{\sigma_0(n)}{n^s}=\zeta(s)\zeta(s-1).$$

For $s > 2$, and more generally,

$$\sum_{n=1}^{\infty} \frac{\sigma_0(n)}{n^s} = \zeta(s)\zeta(s-k).$$

For $s > 1$, and $k \geq 0$.

A generating function for $\sigma_0(n)$ is given by the Lambert series,

$$\begin{aligned} L(x) &= \sum_{n=1}^{\infty} \frac{x^n}{1-x^n} \\ &= \frac{\psi_x(1) + \text{In}(1-x)}{\text{In}\, x} \\ &= \sigma_0(1)x + \sigma_0(2)x^2 + \ldots \\ &= x + 2x^2 + 2x^3 + 3x^4 + 2x^5 + \ldots, \end{aligned}$$

Where $\phi_q(x)$ is a q-polygamma function.

The $\sigma_1(n)$ function has the series expansion:

$$\sigma_1(n) = \frac{1}{6} n\pi^2 \left[\left(1 + \frac{(-1)^n}{2^2}\right) + \frac{2\cos\left(\frac{2}{3} n\pi\right)}{3^2} + \frac{2\cos\left(\frac{1}{2} n\pi\right)}{4^2} + \frac{2\left[\cos\left(\frac{2}{5} n\pi\right) + \cos\left(\frac{4}{5} n\pi\right)\right]}{5^2} + \ldots \right].$$

Ramanujan gave the beautiful formula:

$$\sum_{n=1}^{\infty} \frac{\sigma_a(n)\sigma_b(n)}{n^s} = \frac{\zeta(s)\zeta(s-a)\zeta(s-b)\zeta(s-a-b)}{\zeta(2s-a-b)}.$$

Where $\zeta(n)$ is the zeta function and $R[s], R[s-a], R[s-b], R[s-a-b] > 1$, which was used by Ingham in a proof of the prime number theorem. This gives the special case:

$$\sum_{n=1}^{\infty} \frac{[d(n)]^2}{n^s} = \frac{[\zeta(s)]^4}{\zeta(2s)}.$$

Gronwall's theorem states that:

$$\overline{\lim_{n \to \infty}} \frac{\sigma_1(n)}{n\, \text{In}\, \text{In}\, n} = e^{\tilde{a}}.$$

Where $\tilde{a}$ is the Euler-Mascheroni constant. This can be written as an explicit inequality as:

$$\frac{\sigma_1(n)}{n\, \text{In}\, \text{In}\, n} \leq e^{\tilde{a}} + \frac{c}{(\text{In}\, \text{In}\, n)^2}.$$

where $\tilde{a}$ is the Euler-Mascheroni constant and where equality holds for $n = 12$, giving:

$$c = \frac{7}{3}(\text{In In}12) - e^{\tilde{a}}(\text{In In}12)^2$$
$$= 0.648213649...$$

In fact, the constant term can be dropped if the Riemann hypothesis holds, since the Riemann hypothesis is equivalent to the statement that:

$$\frac{\sigma_1(n)}{n \,\text{In}\,\text{In}\, n} < e^{\tilde{a}}.$$

For all $n \geq 5041$.

$\sigma_1(n)$ is a power of 2 iff $n = 1$ or n is a product of distinct Mersenne primes. The first few such n are 1, 3, 7, 21, 31, 93, 127, 217, 381, 651, 889, 2667, ..., and the powers of 2 these correspond to are 0, 2, 3, 5, 5, 7, 7, 8, 9, 10, 10, 12, 12, 13, 14, ...

Curious identities derived using modular form theory are given by,

$$\sigma_3(n) - \sigma_7(n) + 120\sum_{k=1}^{n-1}\sigma_3(k)\sigma_3(n-k) = 0$$

$$-10\sigma_3(n) + 21\sigma_5(n) - 11\sigma_9(n) + 5040\sum_{k=1}^{n-1}\sigma_3(k)\sigma_5(n-k) = 0.$$

Together with,

$$21\sigma_5(n) - 20\sigma_7(n) - \sigma_{13}(n) + 10080\sum_{k=1}^{n-1}\sigma_5(n-k)\sigma_7(k) = 0$$

$$-10\sigma_3(n) + 11\sigma_9(n) - \sigma_{13}(n) + 2640\sum_{k=1}^{n-1}\sigma_3(n-k)\sigma_9(k) = 0$$

$$-21\sigma_5(n) + 22\sigma_9(n) - \sigma_{13}(n) - 2904\sum_{k=1}^{n-1}\sigma_9(n-k)\sigma_9(k)$$

$$+504\sum_{k=1}^{n-1}\sigma_5(n-k)\sigma_{13}(k) = 0.$$

The divisor function $\sigma_1(n)$ (and, in fact, $\sigma_k(n)$ for $k \geq 1$) is odd iff n is a square number or twice a square number. The divisor function $\sigma_1(n)$ satisfies the congruence:

$$n\sigma_1(n) \equiv 2(\text{mod}\,\phi(n)),$$

for all primes and no composite numbers with the exception of 4, 6, and 22.

The number of divisors $d(n)$ is prime whenever $\sigma_1(n)$ itself is prime. Factorizations of $\sigma_1(p^a)$ for prime p are given by Sorli.

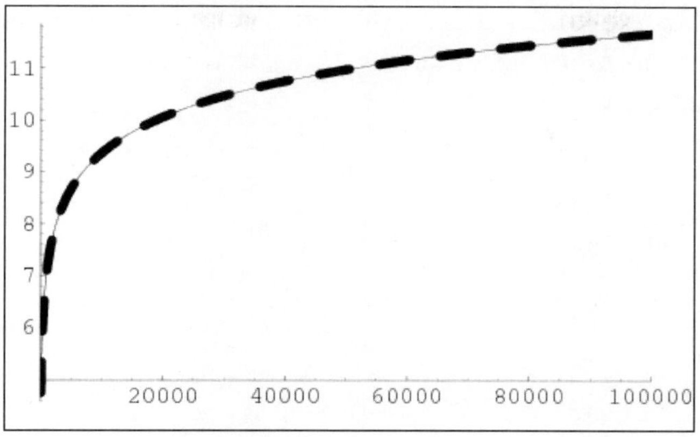

In 1838, Dirichlet showed that the average number of divisors of all numbers from 1 to n is asymptotic to,

$$\frac{\sum_{k=1}^{n} d(k)}{n} \sim \mathrm{In}\, n + 2\gamma - 1.$$

As illustrated above, where the thin solid curve plots the actual values and the thick dashed curve plots the asymptotic function. This is related to the Dirichlet divisor problem, which seeks to find the "best" coefficient θ in,

$$\sum_{k=1}^{n} d(k) = n\,\mathrm{In}\, n + (2\gamma - 1)n + O(\eta^{\theta})$$

The summatory functions for σ_a with $a > 1$ are,

$$\sum_{k=1}^{n} \sigma_a(k) = \frac{\zeta(a+1)}{a+1} n^{a+1} + O(n^a).$$

For $a = 1$,

$$\sum_{k=1}^{n} \sigma_1(k) = \frac{\pi^2}{12} n^2 + O(n\,\mathrm{In}\; n)$$

The divisor function can also be generalized to Gaussian integers. The definition requires some care since in principle, there is ambiguity as to which of the four associates is chosen for each divisor. Spira defines the sum of divisors of a complex number z by factoring z into a product of powers of distinct Gaussian primes,

$$z = \in \prod {}^{k_i}_{p_i},$$

Where $\in$ is a unit and each p_i lies in the first quadrant of the complex plane, and then writing:

$$\sigma_1(z) = \prod \frac{p_i^{k_i+1} - 1}{p_i - 1}.$$

This makes σ a multiplicative function and also gives $|\sigma_1(z)| \geq z$. This extension is implemented in the Wolfram Language as DivisorSigma[1, z, GaussianIntegers -> True]. The following table gives $\sigma_1(a+ib)$ nonnegative values of a and b.

$a \backslash b$	0	1	2	3	4	5	6
1	1	$2+i$	$2+2i$	$5+5i$	$2+4i$	$6+8i$	$2+6i$
2	$2+3i$	$3+i$	$5i$	$3+3i$	$-2+10i$	$3+5i$	$-5+15i$
3	4	$2+6i$	$4+2i$	$8+4i$	$6+5i$	$9+7i$	$8+8i$
4	$-4+5i$	$5+i$	$3+11i$	$-1+6i$	$-8+i$	$5+5i$	$-3+15i$
5	$4+8i$	$3+9i$	$6+2i$	$10i$	$6+4i$	$20i$	$6+6i$
6	$8+12i$	$7+i$	$-10+10$	$12+4i$	$2+16i$	$7+5i$	$20i$

DIRICHLET CONVOLUTION

In mathematics, the Dirichlet convolution is a binary operation defined for arithmetic functions; it is important in number theory. It was developed by Peter Gustav Lejeune Dirichlet.

If $f, g : \mathbb{N} \to \mathbb{C}$ are two arithmetic functions from the positive integers to the complex numbers, the *Dirichlet convolution* $f * g$ is a new arithmetic function defined by:

$$(f * g)(n) = \sum_{d|n} f(d) g\left(\frac{n}{d}\right) = \sum_{ab=n} f(a) g(b),$$

where the sum extends over all positive divisors d of n, or equivalently over all distinct pairs (a, b) of positive integers whose product is n.

This product occurs naturally in the study of Dirichlet series such as the Riemann zeta function. It describes the multiplication of two Dirichlet series in terms of their coefficients:

$$\left(\sum_{n\geq 1} \frac{f(n)}{n^s}\right)\left(\sum_{n\geq 1} \frac{g(n)}{n^s}\right) = \left(\sum_{n\geq 1} \frac{(f * g)(n)}{n^s}\right).$$

Properties

The set of arithmetic functions forms a commutative ring, the Dirichlet ring, under pointwise addition, where $f + g$ is defined by $(f + g)(n) = f(n) + g(n)$, and Dirichlet convolution. The multiplicative identity is the unit function ε defined by $\varepsilon(n) = 1$ if $n = 1$ and $\varepsilon(n) = 0$ if $n > 1$. The units (invertible elements) of this ring are the arithmetic functions f with $f(1) \neq 0$.

Specifically, Dirichlet convolution is associative,

$$(f*g)*h = f*(g*h),$$

distributes over addition,

$$f*(g+h) = f*g + f*h,$$

is commutative,

$$f*g = g*f,$$

and has an identity element,

$$f*\varepsilon = \varepsilon*f = f.$$

Furthermore, for each f having $f(1) \neq 0$, there exists an arithmetic function f^{-1} with $f*f^{-1} = \varepsilon$, called the Dirichlet inverse of f.

The Dirichlet convolution of two multiplicative functions is again multiplicative, and every not constantly zero multiplicative function has a Dirichlet inverse which is also multiplicative. This means that the subring of the Dirichlet ring composed of the multiplicative functions is also a field. The article on multiplicative functions lists several convolution relations among important multiplicative functions.

Another operation on arithmetic functions is pointwise multiplication: *fg* is defined by (*fg*)(*n*) = *f*(*n*) *g*(*n*). Given a completely multiplicative function h, pointwise multiplication by distributes over Dirichlet convolution: $(f*g)h = (fh)*(gh)$. The convolution of two completely multiplicative functions is multiplicative, but not necessarily completely multiplicative.

Examples:

In these formulas, we use the following arithmetical functions:

- ε is the multiplicative identity: $\varepsilon(1) = 1$,, otherwise 0.
- 1 is the constant function with value 1: $1(n) = 1$ for all n. Keep in mind that 1 is not the identity. (Some authors denote this as ζ because the associated Dirichlet series is the Riemann zeta function).
- 1_C for $C \subset \mathbb{N}$ is a set indicator function: $1_C(n) = 1$ iff $n \in C$, otherwise 0.
- Id is the identity function with value *n*: $\mathrm{Id}(n) = n$.
- Id_k is the *k*th power function: $\mathrm{Id}_k(n) = n^k$.

The following relations hold:

- $1*\mu = \varepsilon$, the Dirichlet inverse of the constant function 1 is the Möbius function.
- $g = f*1$ if and only if $f = g*\mu$, the Möbius inversion formula.
- $\sigma_k = \mathrm{Id}_k * 1$, the kth-power-of-divisors sum function σ_k.

- $\sigma = \text{Id} * 1$, the sum-of-divisors function $\sigma = \sigma_1$.
- $d = 1 * 1$, the number-of-divisors function $d(n) = \sigma_0$.
- $\text{Id}_k = \sigma_k * \mu$, by Möbius inversion of the formulas for σ_k, σ, and d.
- $\text{Id} = \sigma * \mu$.
- $1 = d * \mu$.
- $\phi * 1 = \text{Id}$, proved under Euler's totient function.
- $\phi = \text{Id} * \mu$, by Möbius inversion.
- $\sigma = \phi * d$, from convolving 1 on both sides of $\phi * 1 = \text{Id}$.
- $\lambda * |\mu| = \varepsilon$ where λ is Liouville's function.
- $\lambda * 1 = 1_{Sq}$ where Sq = {1, 4, 9, ...} is the set of squares.
- $\text{Id}_k * (\text{Id}_k \mu) = \varepsilon$.
- $d^3 * 1 = (d * 1)^2$.
- $J_k * 1 = \text{Id}_k$, Jordan's totient function.
- $(Id_s J_r) * J_s = J_{s+r}$.
- $\Lambda * 1 = \log$, where Λ is von Mangoldt's function.
- $|\mu| * 1 = 2^{\omega}$, where $\omega(n)$ is the prime omega function counting *distinct* prime factors of n.
- $\Omega * \mu = \chi_{2\mathcal{P}}$ is an indicator function where the set $2\mathcal{P} := \{n \geq 1 : n = 2^k \vee n \in \mathbb{P}\}$ is the collection of positive primes and integral powers of two.
- $\omega * \mu = \chi_{\mathbb{P}}$ where $\chi_{\mathbb{P}}(n) \mapsto \{0,1\}$ is the characteristic function of the primes.

This last identity shows that the prime counting function is given by the summatory function $\pi(x) = \sum_{n \leq x} (\omega * \mu)(n) = \sum_{d=1}^{x} \omega(d) M\left(\left\lfloor \frac{x}{d} \right\rfloor\right)$ where $M(x)$ is the Mertens function and ω is the distinct prime factor counting function from above. This expansion follows from the identity for the sums over Dirichlet concolutions given on the divisor sum identities page (a standard trick for these sums).

Dirichlet Inverse

Examples:

Given an arithmetic function f its Dirichlet inverse $g = f^{-1}$ may be calculated recursively: the value of $g(n)$ is in terms of $g(m)$ for $m < n$.

For $n=1$:

$(f*g)(1) = f(1)g(1) = \varepsilon(1) = 1$, so

$g(1) = 1/f(1)$. This implies that f does not have a Dirichlet inverse if $f(1) = 0$.

For $n=2$:

$$(f*g)(2) = f(1)g(2) + f(2)g(1) = \varepsilon(2) = 0,$$

$$g(2) = -1/f(1)(f(2)g(1)).$$

For $n=3$:

$$(f*g)(3) = f(1)g(3) + f(3)g(1) = \varepsilon(3) = 0,$$

$$g(3) = -1/f(1)(f(3)g(1)).$$

For $n=4$:

$$(f*g)(4) = f(1)g(4) + f(2)g(2) + f(4)g(1) = \varepsilon(4) = 0,$$

$$g(4) = -1/f(1)(f(4)g(1) + f(2)g(2)).$$

And in general for $n>1$,

$$g(n) = \frac{-1}{f(1)} \sum_{\substack{d|n \\ d<n}} f\left(\frac{n}{d}\right) g(d).$$

Properties

The following properties of the Dirichlet inverse hold:

- The function *f* has a Dirichlet inverse if and only if *f*(1) ≠ 0.
- The Dirichlet inverse of a multiplicative function is again multiplicative.
- The Dirichlet inverse of a Dirichlet convolution is the convolution of the inverses of each function: $(f*g)^{-1} = f^{-1} * g^{-1}$.
- A multiplicative function *f* is completely multiplicative if and only if $f^{-1}(n) = \mu(n) f(n)$.
- If *f* is completely multiplicative then $(f \cdot g)^{-1} = f \cdot g^{-1}$ whenever $g(1) \neq 0$ and where denotes pointwise multiplication of functions.

Other Formulas

Arithmetic function	Dirichlet inverse:
Constant function with value 1	Möbius function μ

n^α	$\mu(n)n^\alpha$
Liouville's function λ	Absolute value of Möbius function |μ|
Euler's totient function φ	$\sum_{d\mid n} d\,\mu(d)$
The generalized sum-of-divisors function σ_α	$\sum_{d\mid n} d^\alpha \mu(d)\mu\left(\frac{n}{d}\right)$

An exact, non-recursive formula for the Dirichlet inverse of any arithmetic function *f* is given in Divisor sum identities. A more partition theoretic expression for the Dirichlet inverse of *f* is given by,

$$f^{-1}(n) = \sum_{k=1}^{\Omega(n)} \left\{ \sum_{\substack{\lambda_1+2\lambda_2+\cdots+k\lambda_k=n \\ \lambda_1,\lambda_2,\ldots,\lambda_k \mid n}} \frac{(\lambda_1+\lambda_2+\cdots+\lambda_k)!}{1!2!\cdots k!} (-1)^k f(\lambda_1) f(\lambda_2)^2 \cdots f(\lambda_k)^k \right\}.$$

Dirichlet Series

If *f* is an arithmetic function, one defines its Dirichlet series generating function by,

$$DG(f;s) = \sum_{n=1}^{\infty} \frac{f(n)}{n^s}.$$

For those complex arguments *s* for which the series converges (if there are any). The multiplication of Dirichlet series is compatible with Dirichlet convolution in the following sense:

$$DG(f;s)DG(g;s) = DG(f*g;s).$$

For all *s* for which both series of the left hand side converge, one of them at least converging absolutely (note that simple convergence of both series of the left hand side DOES NOT imply convergence of the right hand side!). This is akin to the convolution theorem if one thinks of Dirichlet series as a Fourier transform.

Related Concepts

The restriction of the divisors in the convolution to unitary, bi-unitary or infinitary divisors defines similar commutative operations which share many features with the Dirichlet convolution (existence of a Möbius inversion, persistence of multiplicativity, definitions of totients, Euler-type product formulas over associated primes, etc.).

Dirichlet convolution is the convolution of the incidence algebra for the positive integers ordered by divisibility.

LIOUVILLE FUNCTION

The Liouville function, denoted by $\lambda(n)$ and named after Joseph Liouville, is an important function in number theory.

If n is a positive integer, then $\lambda(n)$ is defined as:

$$\lambda(n) = (-1)^{\Omega(n)},$$

where $\Omega(n)$ is the number of prime factors of n, counted with multiplicity. If n is squarefree, i.e., if $n = p_1 p_2 \cdots p_k$ where p_i is prime for all i and where $p_i \neq p_j \forall i \neq j$, then we have the following alternate formula for the function expressed in terms of the Möbius function and the distinct prime factor counting function $\omega(n)$:

$$\lambda(n) = \mu(n) = \mu^2(n)(-1)^{\omega(n)}.$$

λ is completely multiplicative since $\Omega(n)$ is completely additive, i.e.: $\Omega(ab) = \Omega(a) + \Omega(b)$. The number 1 has no prime factors, so $\Omega(1) = 0$ and therefore $\lambda(1) = 1$. The Liouville function satisfies the identity:

$$\sum_{d|n} \lambda(d) = \begin{cases} 1 & \text{if } n \text{ is a perfect square,} \\ 0 & \text{otherwise.} \end{cases}$$

The Dirichlet inverse of Liouville function is the absolute value of the Möbius function, $\lambda^{-1}(n) = |\mu(n)| = \mu^2(n)$, which is equivalently the characteristic function of the squarefree integers. We also have that $\lambda(n)\mu(n) = \mu^2(n)$, and that for all natural numbers n:

$$\lambda(n) = \sum_{d^2|n} \mu\left(\frac{n}{d^2}\right).$$

Series

The Dirichlet series for the Liouville function is related to the Riemann zeta function by:

$$\frac{\zeta(2s)}{\zeta(s)} = \sum_{n=1}^{\infty} \frac{\lambda(n)}{n^s}.$$

The Lambert series for the Liouville function is:

$$\sum_{n=1}^{\infty} \frac{\lambda(n) q^n}{1-q^n} = \sum_{n=1}^{\infty} q^{n^2} = \frac{1}{2}\left(\vartheta_3(q) - 1\right),$$

where $\vartheta_3(q)$ is the Jacobi theta function.

Conjectures on Weighted Summatory Functions

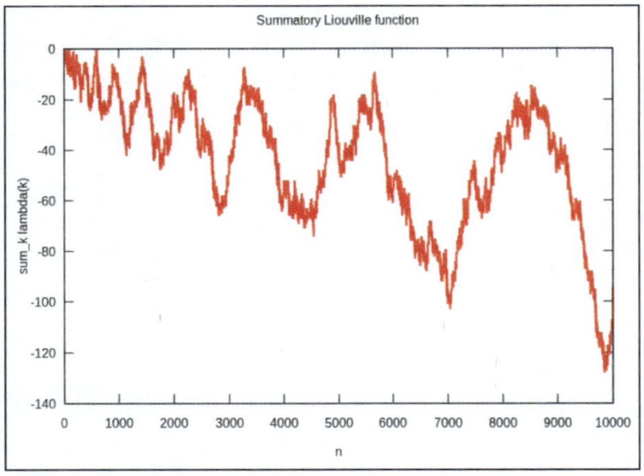

Summatory Liouville function L(n) up to n = 10^4. The readily visible oscillations are due to the first non-trivial zero of the Riemann zeta function.

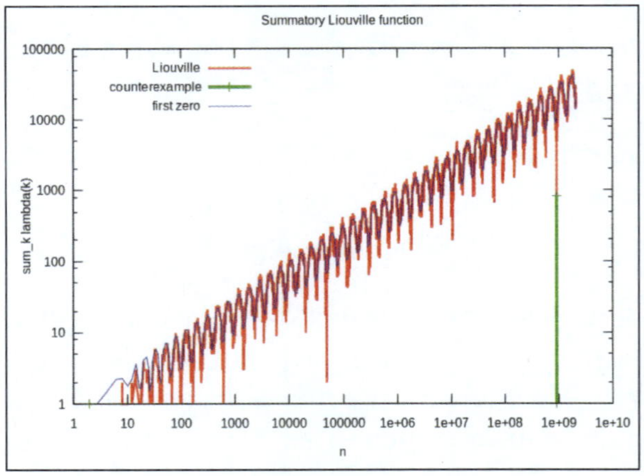

Logarithmic graph of the negative of the summatory Liouville function L(n) up to n = 2×10^9.

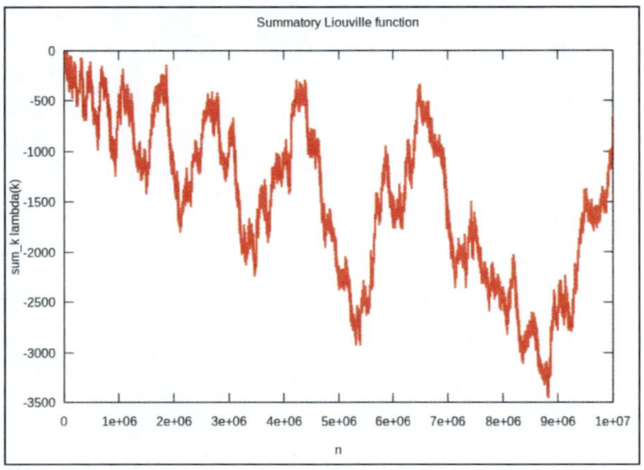

Summatory Liouville function L(n) up to n = 10^7. Note the apparent scale invariance of the oscillations.

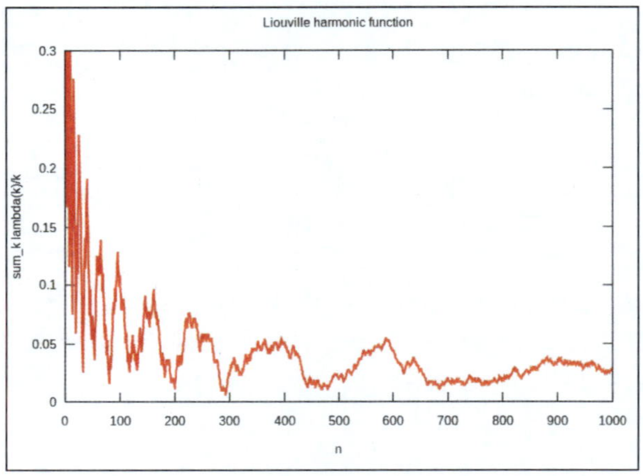

Harmonic Summatory Liouville function T(n) up to n = 10^3.

The Pólya conjecture is a conjecture made by George Pólya in 1919. Defining:

$$L(n) = \sum_{k=1}^{n} \lambda(k),$$

the conjecture states that $L(n) \leq 0$ for $n > 1$. This turned out to be false. The smallest counter-example is $n = 906150257$, found by Minoru Tanaka in 1980. It has since been shown that $L(n) > 0.0618672\sqrt{n}$ for infinitely many positive integers n, while it can also be shown via the same methods that $L(n) < -1.3892783\sqrt{n}$ for infinitely many positive integers n.

For any $\varepsilon > 0$, assuming the Riemann hypothesis, we have that the summatory function $L(x) \equiv L_0(x)$ is bounded by,

$$L(x) = O\left(\sqrt{x}\exp\left(C \cdot \log^{1/2}(x)\left(\log\log x\right)^{5/2+\varepsilon}\right)\right),$$

where the $C > 0$ is some absolute limiting constant.

Define the related sum:

$$T(n) = \sum_{k=1}^{n} \frac{\lambda(k)}{k}.$$

It was open for some time whether $T(n) \geq 0$ for sufficiently big $n \geq n_0$ (this conjecture is occasionally–though incorrectly–attributed to Pál Turán). This was then disproved by Haselgrove, who showed that $T(n)$ takes negative values infinitely often. A confirmation of this positivity conjecture would have led to a proof of the Riemann hypothesis, as was shown by Pál Turán.

Generalizations

More generally, we can consider the weighted summatory functions over the Lioville function defined for any $\alpha \in \mathbb{R}$ as follows for positive integers x where we have the special cases $L(x) := L_0(x)$ and $T(x) = L_1(x)$

$$L_\alpha(x) := \sum_{n \leq x} \frac{\lambda(n)}{n^\alpha}.$$

These α^{-1} -weighted summatory functions are related to the Mertens function, or weighted summatory functions of the Moebius function. In fact, we have that the so-termed non-weigthed, or ordinary function $L(x)$ precisely corresponds to the sum

$$L(x) = \sum_{d^2 \leq x} M\left(\frac{x}{d^2}\right) = \sum_{d^2 \leq x} \sum_{n \leq \frac{x}{d^2}} \mu(n).$$

Moreover, as noted in these functions satisfy similar bounding asymptotic relations. For example, whenever $0 \leq \alpha \leq \frac{1}{2}$, we see that there exists an absolute constant $C_\alpha > 0$ such that:

$$L_\alpha(x) = O\left(x^{1-\alpha} \exp\left(-C_\alpha \frac{(\log x)^{3/5}}{(\log \log x)^{1/5}}\right)\right).$$

By an application of Perron's formula, or equivalently by a key (inverse) Mellin transform, we have that:

$$\frac{\zeta(2\alpha + 2s)}{\zeta(\alpha + s)} = s \cdot \int_1^\infty \frac{L_\alpha(x)}{x^{s+1}} dx,$$

which then can be inverted via the inverse transform to show that for $x > 1$, $T \geq 1$ and $0 \leq \alpha < \frac{1}{2}$.

$$L_\alpha(x) = \frac{1}{2\pi\iota} \int_{\sigma_0 - \iota T}^{\sigma_0 + \iota T} \frac{\zeta(2\alpha + 2s)}{\zeta(\alpha + s)} \cdot \frac{x^s}{s} ds + E_\alpha(x) + R_\alpha(x, T),$$

where we can take $\sigma_0 := 1-\alpha+1/\log(x)$, and with the remainder terms defined such that $E_\alpha(x) = O(x^{-\alpha})$ and $R_\alpha(x,T) \to 0$ as $T \to \infty$.

In particular, if we assume that the Riemann hypothesis (RH) is true and that all of the non-trivial zeros, denoted by $\rho = \frac{1}{2} + \imath\gamma$, of the Riemann zeta function are simple, then for any $0 \le \alpha < \frac{1}{2}$ and $x \ge 1$ there exists an infinite sequence of $\{T_v\}_{v\ge 1}$ which satisfies that $v \le T_v \le v+1$ for all v such that:

$$L_\alpha(x) = \frac{x^{1/2-\alpha}}{(1-2\alpha)\zeta(1/2)} + \sum_{|\gamma|<T_v} \frac{\zeta(2\rho)}{\zeta'(\rho)} \cdot \frac{x^{\rho-\alpha}}{(\rho-\alpha)} + E_\alpha(x) + R_\alpha(x,T_v) + I_\alpha(x),$$

where for any increasingly small $0 < \varepsilon < \frac{1}{2} - \alpha$ we define:

$$I_\alpha(x) := \frac{1}{2\pi \acute{y} x^\alpha} \int_{\varepsilon+\alpha-\imath\infty}^{\varepsilon+\alpha+\imath\infty} \frac{\zeta(2s)}{\zeta(s)} \cdot \frac{x^s}{(s-\alpha)} ds,$$

and where the remainder term:

$$R_\alpha(x,T) \ll x^{-\alpha} + \frac{x^{1-\alpha}\log(x)}{T} + \frac{x^{1-\alpha}}{T^{1-\varepsilon}\log(x)},$$

which of course tends to *o* as $T \to \infty$. These exact analytic formula expansions again share similar properties to those corresponding to the weighted Mertens function cases. Additionally, since $\zeta(1/2) < 0$ we have another similarity in the form of $L_\alpha(x)$ to $M(x)$ in so much as the dominant leading term in the previous formulas predicts a negative bias in the values of these functions over the positive natural numbers *x*.

PARTITION FUNCTION

In number theory, the partition function $p(n)$ represents the number of possible partitions of a non-negative integer n. For instance, $p(4) = 5$ because the integer 4 has the five partitions $1+1+1+1$, $1+1+2$, $1+3$, $2+2$, and 4.

No closed-form expression for the partition function is known, but it has both asymptotic expansions that accurately approximate it and recurrence relations by which it can be calculated exactly. It grows as an exponential function of the square root of its argument. The multiplicative inverse of its generating function is the Euler function; by Euler's pentagonal number theorem this function is an alternating sum of pentagonal number powers of its argument.

Srinivasa Ramanujan first discovered that the partition function has nontrivial patterns in modular arithmetic, now known as Ramanujan's congruences. For instance, whenever the decimal representation of n ends in the digit 4 or 9, the number of partitions of n will be divisible by 5.

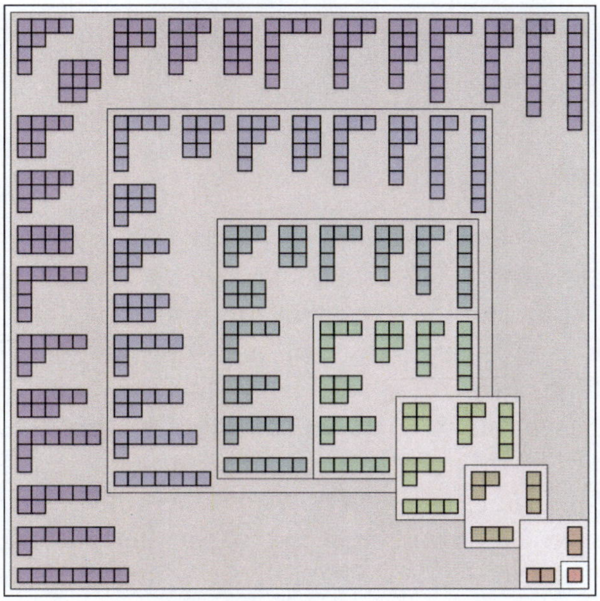

The values $p(1),\ldots p(8)$ of the partition function (1, 2, 3, 5, 7, 11, 15, and 22) can be determined by counting the Young diagrams for the partitions of the numbers from 1 to 8.

For a positive integer n, $p(n)$ is the number of distinct ways of representing n as a sum of positive integers. For the purposes of this definition, the order of the terms in the sum is irrelevant: two sums with the same terms in a different order are not considered to be distinct.

By convention $p(0)=1$, as there is one way (the empty sum) of representing zero as a sum of positive integers. For the same reason, by definition, $p(n)=0$ when n is negative.

The first few values of the partition function, starting with $p(0)=1$, are:

1, 1, 2, 3, 5, 7, 11, 15, 22, 30, 42, 56, 77, 101, 135, 176, 231, 297, 385, 490, 627, 792, 1002, 1255, 1575, 1958, 2436, 3010, 3718, 4565, 5604, ...

Some exact value of $p(n)$ for larger values of n include:

$p(100)=190{,}569{,}292.$

$p(1000)=24{,}061{,}467{,}864{,}032{,}622{,}473{,}692{,}149{,}727{,}991 \approx 2.40615\times10^{31}.$

$p(10000)=36{,}167{,}251{,}325{,}\ldots{,}906{,}916{,}435{,}144 \approx 3.61673\times10^{106}.$

As of September 2017, the largest known prime number among the values of $p(n)$ is $p(221444161)$, with $16{,}569$ decimal digits.

Generating Function

The generating function for $p(n)$ is given by,

$$\sum_{n=0}^{\infty} p(n)x^n = \prod_{k=1}^{\infty}\left(\frac{1}{1-x^k}\right)$$

$$= (1 + x + x^2 + x^3 + \cdots)(1 + x^2 + x^4 + x^6 + \cdots)(1 + x^3 + x^6 + x^9 + \cdots)\cdots$$
$$= \frac{1}{1 - x - x^2 + x^5 + x^7 - x^{12} - x^{15} + x^{22} + x^{26} - \cdots}$$
$$= 1 \Big/ \sum_{k=-\infty}^{\infty} (-1)^k x^{k(3k-1)/2}.$$

The equality between the products on the first and second lines of this formula is obtained by expanding each factor $1/(1-x^k)$ into the geometric series $(1 + x^k + x^{2k} + x^{3k} + \cdots)$. To see that the expanded product equals the sum on the first line, apply the distributive law to the product. This expands the product into a sum of monomials of the form $x^{a_1} x^{2a_2} x^{3a_3} \cdots$ for some sequence of coefficients a_i, only finitely many of which can be non-zero. The exponent of the term is $n = \sum ia_i$, and this sum can be interpreted as a representation of as a partition into a_i copies of each number i. Therefore, the number of terms of the product that that have exponent n is exactly $p(n)$, the same as the coefficient of x^n in the sum on the left. Therefore, the sum equals the product.

The function that appears in the denominator in the third and fourth lines of the formula is the Euler function. The equality between the product on the first line and the formulas in the third and fourth lines is Euler's pentagonal number theorem. The exponents of x in these lines are the pentagonal numbers $P_k = k(3k-1)/2$ for $k \in \{0, 1, -1, 2, -2, \ldots\}$ (generalized somewhat from the usual pentagonal numbers, which come from the same formula for the positive values of k). The pattern of positive and negative signs in the third line comes from the term $(-1)^k$ in the fourth line: even choices of k produce positive terms, and odd choices produce negative terms.

More generally, the generating function for the partitions of n into numbers selected from a set A of positive integers can be found by taking only those terms in the first product for which $k \in A$. This result is due to Leonhard Euler. The formulation of Euler's generating function is a special case of a q-Pochhammer symbol and is similar to the product formulation of many modular forms, and specifically the Dedekind eta function.

Recurrence Relations

The same sequence of pentagonal numbers appears in a recurrence relation for the partition function:

$$p(n) = \sum_{k \neq 0} (-1)^{k+1} p(n - k(3k-1)/2)$$
$$= p(n-1) + p(n-2) - p(n-5) - p(n-7) + p(n-12) + p(n-15) - p(n-22) - \cdots$$

As base cases, $p(0)$ is taken to equal 1, and $p(k)$ is taken to be zero for negative k. Although the sum on the right side appears infinite, it has only finitely many nonzero terms, coming from the nonzero values of k in the range:

$$-\frac{\sqrt{24n+1}-1}{6} \le k \le \frac{\sqrt{24n+1}+1}{6}.$$

Another recurrence relation for $p(n)$ can be given in terms of the sum of divisors function σ:

$$p(n) = \frac{1}{n}\sum_{k=0}^{n-1} \sigma(n-k)p(k).$$

If $q(n)$ denotes the number of partitions of n with no repeated parts then it follows by splitting each partition into its even parts and odd parts, and dividing the even parts by two, that:

$$p(n) = \sum_{k=0}^{\lfloor n/2 \rfloor} q(n-2k)p(k).$$

Congruences

Srinivasa Ramanujan is credited with discovering that the partition function has nontrivial patterns in modular arithmetic. For instance the number of partitions is divisible by five whenever the decimal representation of n ends in the digit 4 or 9, as expressed by the congruence:

$$p(5k+4) \equiv 0 \pmod 5.$$

For instance, the number of partitions for the integer 4 is 5. For the integer 9, the number of partitions is 30; for 14 there are 135 partitions. This congruence is implied by the more general identity:

$$\sum_{k=0}^{\infty} p(5k+4)x^k = 5\frac{(x^5)_\infty^5}{(x)_\infty^6},$$

also by Ramanujan, where the notation $(x)_\infty$ denotes the product defined by:

$$(x)_\infty = \prod_{m=1}^{\infty}(1-x^m).$$

A short proof of this result can be obtained from the partition function generating function.

Ramanujan also discovered congruences modulo 7 and 11:

$$p(7k+5) \equiv 0 \pmod 7,$$
$$p(11k+6) \equiv 0 \pmod{11}.$$

They come from Ramanujan's identity:

$$\sum_{k=0}^{\infty} p(7k+5)x^k = 7\frac{(x^7)_\infty^3}{(x)_\infty^4} + 49x\frac{(x^7)_\infty^7}{(x)_\infty^8}.$$

Since 5, 7, and 11 are consecutive primes, one might think that there would be an analogous congruence for the next prime 13, $p(13k+a) \equiv 0 \pmod{13}$ for some a. However, there is no congruence of the form $p(bk+a) \equiv 0 \pmod b$ for any prime b other than 5, 7, or 11. Instead, to obtain a congruence, the argument of p should take the form $cbk+a$ for some $c > 1$. In the 1960s, A. O. L.

Atkin of the University of Illinois at Chicago discovered additional congruences of this form for small prime moduli. For example:

$$p(11^3 \cdot 13 \cdot k + 237) \equiv 0 \pmod{13}.$$

Ken Ono proved that there are such congruences for every prime modulus greater than 3. Later, Ahlgren & Ono showed there are partition congruences modulo every integer coprime to 6.

Approximation Formulas

Approximation formulas exist that are faster to calculate than the exact formula given above.

An asymptotic expression for $p(n)$ is given by:

$$p(n) \sim \frac{1}{4n\sqrt{3}} \exp\left(\pi\sqrt{\frac{2n}{3}}\right) \text{ as } n \to \infty.$$

This asymptotic formula was first obtained by G. H. Hardy and Ramanujan in 1918 and independently by J. V. Uspensky in 1920. Considering $p(1000)$, the asymptotic formula gives about 2.4402×10^{31}, reasonably close to the exact answer given above (1.415% larger than the true value).

Hardy and Ramanujan obtained an asymptotic expansion with this approximation as the first term:

$$p(n) \sim \frac{1}{2\pi\sqrt{2}} \sum_{k=1}^{v} A_k(n)\sqrt{k} \cdot \frac{d}{dn}\left(\frac{1}{\sqrt{n - \frac{1}{24}}} \exp\left[\frac{\pi}{k}\sqrt{\frac{2}{3}\left(n - \frac{1}{24}\right)} \right] \right),$$

where

$$A_k(n) = \sum_{0 \le m < k, (m,k)=1} e^{\pi i (s(m,k) - 2nm/k)}.$$

Here, the notation $(m,k) = 1$ implies that the sum should occur only over the values of that are relatively prime to k. The function $s(m,k)$ is a Dedekind sum.

The error after v terms is of the order of the next term, and v may be taken to be of the order of $\sqrt{n}$. As an example, Hardy and Ramanujan showed that $p(200)$ is the nearest integer to the sum of the first $v = 5$ terms of the series.

In 1937, Hans Rademacher was able to improve on Hardy and Ramanujan's results by providing a convergent series expression for $p(n)$. It is:

$$p(n) = \frac{1}{\pi\sqrt{2}} \sum_{k=1}^{\infty} A_k(n)\sqrt{k} \cdot \frac{d}{dn}\left(\frac{1}{\sqrt{n - \frac{1}{24}}} \sinh\left[\frac{\pi}{k}\sqrt{\frac{2}{3}\left(n - \frac{1}{24}\right)} \right] \right).$$

The proof of Rademacher's formula involves Ford circles, Farey sequences, modular symmetry and the Dedekind eta function.

It may be shown that the kth term of Rademacher's series is of the order

$$\exp\left(\frac{\pi}{k}\sqrt{\frac{2n}{3}}\right),$$

so that the first term gives the Hardy–Ramanujan asymptotic approximation. Paul Erdős published an elementary proof of the asymptotic formula for $p(n)$.

Partition

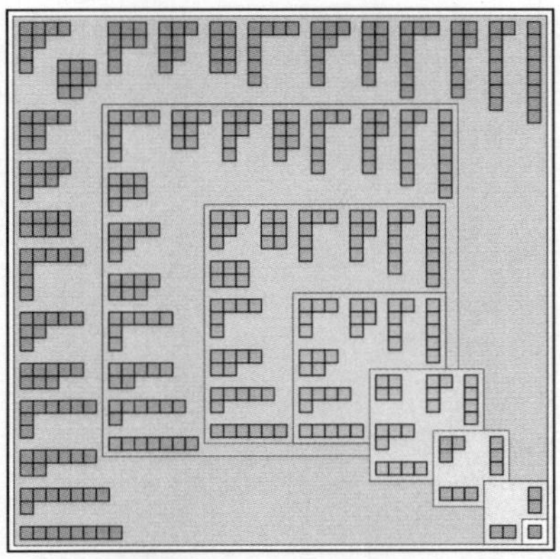

Young diagrams associated to the partitions of the positive integers 1 through 8. They are arranged so that images under the reflection about the main diagonal of the square are conjugate partitions.

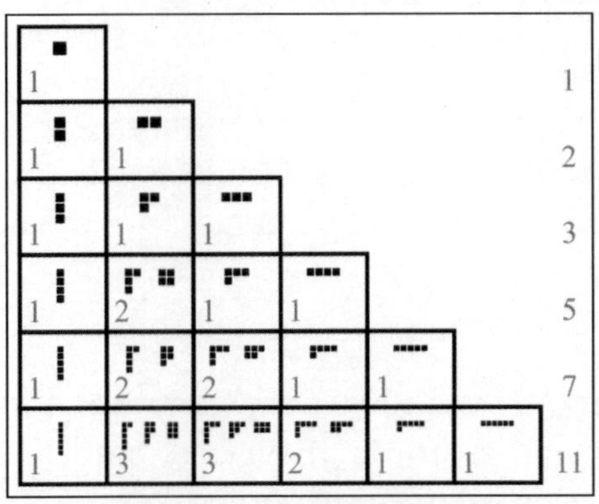

Partitions of n with biggest addend k.

In number theory and combinatorics, a partition of a positive integer n, also called an integer partition, is a way of writing n as a sum of positive integers. Two sums that differ only in the order of their summands are considered the same partition. (If order matters, the sum becomes a composition.) For example, 4 can be partitioned in five distinct ways:

- 4.
- 3 + 1.
- 2 + 2.

- 2 + 1 + 1.
- 1 + 1 + 1 + 1.

The order-dependent composition 1 + 3 is the same partition as 3 + 1, while the two distinct compositions 1 + 2 + 1 and 1 + 1 + 2 represent the same partition 2 + 1 + 1.

A summand in a partition is also called a part. The number of partitions of n is given by the partition function $p(n)$. So $p(4) = 5$. The notation $\lambda \vdash n$ means that λ is a partition of n.

Partitions can be graphically visualized with Young diagrams or Ferrers diagrams. They occur in a number of branches of mathematics and physics, including the study of symmetric polynomials and of the symmetric group and in group representation theory in general.

Examples:

The seven partitions of 5 are:

- 5.
- 4 + 1.
- 3 + 2.
- 3 + 1 + 1.
- 2 + 2 + 1.
- 2 + 1 + 1 + 1.
- 1 + 1 + 1 + 1 + 1.

In some sources partitions are treated as the sequence of summands, rather than as an expression with plus signs. For example, the partition 2 + 2 + 1 might instead be written as the tuple (2, 2, 1) or in the even more compact form $(2^2, 1)$ where the superscript indicates the number of repetitions of a term.

Representations of Partitions

There are two common diagrammatic methods to represent partitions: as Ferrers diagrams, named after Norman Macleod Ferrers, and as Young diagrams, named after the British mathematician Alfred Young. Both have several possible conventions; here, we use *English notation*, with diagrams aligned in the upper-left corner.

Ferrers Diagram

The partition 6 + 4 + 3 + 1 of the positive number 14 can be represented by the following diagram:

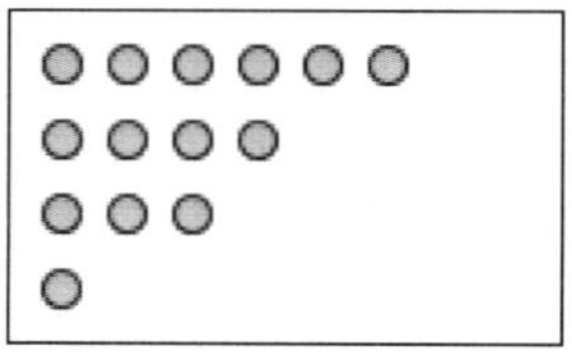

The 14 circles are lined up in 4 rows, each having the size of a part of the partition. The diagrams for the 5 partitions of the number 4 are listed below:

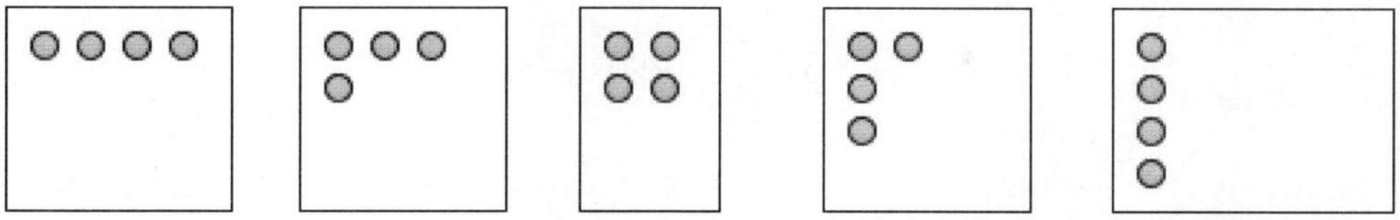

4 = 3 + 1 = 2 + 2 = 2 + 1 + 1 = 1 + 1 + 1 + 1

Young Diagram

An alternative visual representation of an integer partition is its *Young diagram* (often also called a Ferrers diagram). Rather than representing a partition with dots, as in the Ferrers diagram, the Young diagram uses boxes or squares. Thus, the Young diagram for the partition 5 + 4 + 1 is:

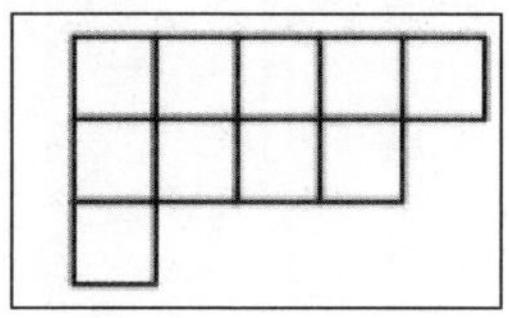

while the Ferrers diagram for the same partition is:

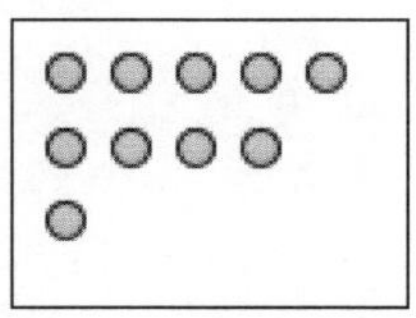

While this seemingly trivial variation doesn't appear worthy of separate mention, Young diagrams turn out to be extremely useful in the study of symmetric functions and group representation theory: filling the boxes of Young diagrams with numbers (or sometimes more complicated objects) obeying various rules leads to a family of objects called Young tableaux, and these tableaux have combinatorial and representation-theoretic significance. As a type of shape made by adjacent squares joined together, Young diagrams are a special kind of polyomino.

Partition Function

The partition function $p(n)$ represents the number of possible partitions of a non-negative integer n. For instance, $p(4)=5$ because the integer 4 has the five partitions $1+1+1+1$, $1+1+2$, $1+3$, $2+2$, and 4. The values of this function for $n=0,1,2...$ are:

1, 1, 2, 3, 5, 7, 11, 15, 22, 30, 42, 56, 77, 101, 135, 176, 231, 297, 385, 490, 627, 792, 1002, 1255, 1575, 1958, 2436, 3010, 3718, 4565, 5604,...

No closed-form expression for the partition function is known, but it has both asymptotic expansions that accurately approximate it and recurrence relations by which it can be calculated exactly. It grows as an exponential function of the square root of its argument. The multiplicative inverse of its generating function is the Euler function; by Euler's pentagonal number theorem this function is an alternating sum of pentagonal number powers of its argument.

Srinivasa Ramanujan first discovered that the partition function has nontrivial patterns in modular arithmetic, now known as Ramanujan's congruences. For instance, whenever the decimal representation of n ends in the digit 4 or 9, the number of partitions of n will be divisible by 5.

Restricted Partitions

In both combinatorics and number theory, families of partitions subject to various restrictions are often studied.

Conjugate and Self-conjugate Partitions

If we flip the diagram of the partition $6 + 4 + 3 + 1$ along its main diagonal, we obtain another partition of 14:

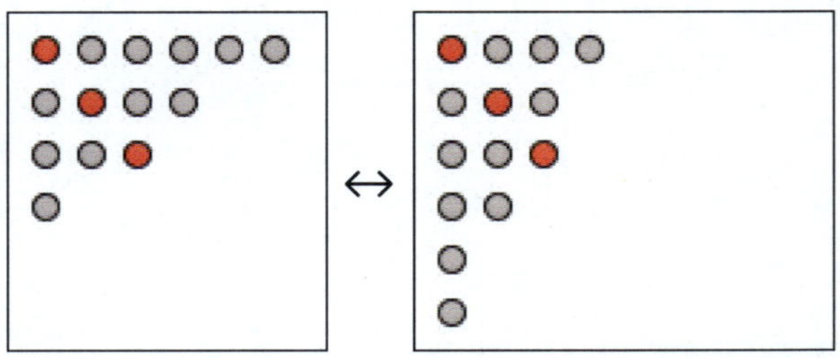

$6 + 4 + 3 + 1 \quad = \quad 4 + 3 + 3 + 2 + 1 + 1$

By turning the rows into columns, we obtain the partition $4 + 3 + 3 + 2 + 1 + 1$ of the number 14. Such partitions are said to be conjugate of one another. In the case of the number 4, partitions 4 and $1 + 1 + 1 + 1$ are conjugate pairs, and partitions $3 + 1$ and $2 + 1 + 1$ are conjugate of each other. Of particular interest is the partition $2 + 2$, which has itself as conjugate. Such a partition is said to be self-conjugate.

Claim: The number of self-conjugate partitions is the same as the number of partitions with distinct odd parts.

Proof (outline): The crucial observation is that every odd part can be "*folded*" in the middle to form a self-conjugate diagram:

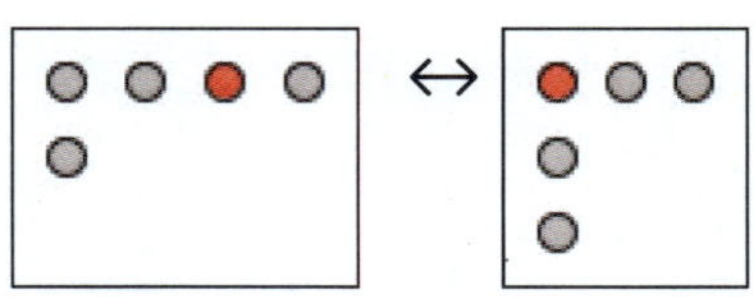

One can then obtain a bijection between the set of partitions with distinct odd parts and the set of self-conjugate partitions, as illustrated by the following example:

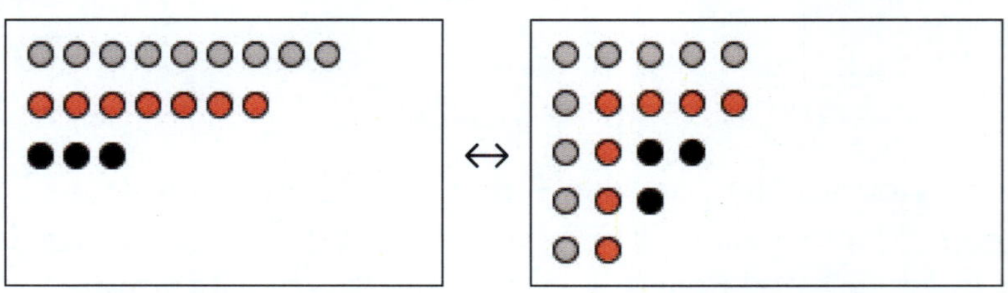

$9 + 7 + 3 \quad = \quad 5 + 5 + 4 + 3 + 2$

Dist. odd self-conjugate

Odd Parts and Distinct Parts

Among the 22 partitions of the number 8, there are 6 that contain only odd parts:

- 7 + 1.
- 5 + 3.
- 5 + 1 + 1 + 1.
- 3 + 3 + 1 + 1.
- 3 + 1 + 1 + 1 + 1 + 1.
- 1 + 1 + 1 + 1 + 1 + 1 + 1 + 1.

Alternatively, we could count partitions in which no number occurs more than once. Such a partition is called a partition with distinct parts. If we count the partitions of 8 with distinct parts, we also obtain 6:

- 8.
- 7 + 1.
- 6 + 2.
- 5 + 3.
- 5 + 2 + 1.
- 4 + 3 + 1.

This is a general property. For each positive number, the number of partitions with odd parts equals the number of partitions with distinct parts, denoted by $q(n)$. This result was proved by Leonhard Euler in 1748 and later was generalized as Glaisher's theorem.

For every type of restricted partition there is a corresponding function for the number of partitions satisfying the given restriction. An important example is $q(n)$. The first few values of $q(n)$ are (starting with $q(0)=1$):

1, 1, 1, 2, 2, 3, 4, 5, 6, 8, 10, ...

The generating function for $q(n)$ (partitions into distinct parts) is given by:

$$\sum_{n=0}^{\infty} q(n)x^n == \prod_{k=1}^{\infty}(1+x^k) = \prod_{k=1}^{\infty}\frac{1}{1-x^{2k-1}}.$$

The pentagonal number theorem gives a recurrence for q:

$$q(k) = a_k + q(k-1) + q(k-2) - q(k-5) - q(k-7) + q(k-12) + q(k-15) - q(k-22) - \ldots$$

where a_k is $(-1)^m$ if $k = 3m^2 - m$ for some integer m and is 0 otherwise.

Restricted Part Size or Number of Parts

By taking conjugates, the number $p_k(n)$ of partitions of n into exactly k parts is equal to the number of partitions of n in which the largest part has size k. The function $p_k(n)$ satisfies the recurrence,

$$p_k(n) = p_k(n-k) + p_{k-1}(n-1)$$

with initial values $p_0(0) = 1$ and $p_k(n) = 0$ if $n \leq 0$ or $k \leq 0$ and n and k are not both zero.

One recovers the function $p(n)$ by:

$$p(n) = \sum_{k=1}^{n} pk(n).$$

One possible generating function for such partitions, taking k fixed and n variable, is,

$$\sum_{n\geq 0} pk(n)x^n = x^k \cdot \prod_{i=1}^{k} \frac{1}{1-x^i}.$$

More generally, if T is a set of positive integers then the number of partitions of n, all of whose parts belong to T, has generating function:

$$\prod_{t\in T}(1-x^t)^{-1}.$$

This can be used to solve change-making problems (where the set T specifies the available coins). As two particular cases, one has that the number of partitions of n in which all parts are 1 or 2 (or, equivalently, the number of partitions of n into 1 or 2 parts) is:

$$\left\lfloor \frac{n}{2}+1 \right\rfloor,$$

and the number of partitions of n in which all parts are 1, 2 or 3 (or, equivalently, the number of partitions of n into at most three parts) is the nearest integer to $(n + 3)^2 / 12$.

Asymptotics

The asymptotic growth rate for $p(n)$ is given by,

$$\log p(n) \sim C\sqrt{n} \text{ as } n \to \infty$$

where $C = \pi\sqrt{\frac{2}{3}}$.

If A is a set of natural numbers, we let $p_A(n)$ denote the number of partitions of n into elements of A. If A possesses positive natural density α then

$$\log p_A(n) \sim C\sqrt{\alpha n}$$

and conversely if this asymptotic property holds for $p_A(n)$ then A has natural density α. This result was stated, with a sketch of proof, by Erdős in 1942.

If A is a finite set, this analysis does not apply (the density of a finite set is zero). If A has k elements whose greatest common divisor is 1, then

$$pA(n) = \left(\prod_{\alpha \in A} a^{-1} \right) \cdot \frac{n^{k-1}}{(k-1)!} + O(n^{k-2}).$$

Partitions in a Rectangle and Gaussian Binomial Coefficients

One may also simultaneously limit the number and size of the parts. Let $p(N, M; n)$ denote the number of partitions of n with at most M parts, each of size at most N. Equivalently, these are the partitions whose Young diagram fits inside an $M \times N$ rectangle. There is a recurrence relation

$$p(n)(N, M; = n) = p(N, M-1; n) + p(N-1, M; n-M)$$

obtained by observing that $p(N, M-; n) - p(N, M-1; n)$ counts the partitions of n into exactly M parts of size at most N, and subtracting 1 from each part of such a partitions yields a partition of $n - M$ into at most M parts.

The Gaussian binomial coefficient is defined as:

$$\binom{k+\ell}{\ell}_q = \binom{k=\ell}{k}_q = \frac{\prod_{j=1}^{k=\ell}(1-q^j)}{\prod_{j=1}^{k}(1-q^j)\prod_{j=1}^{\ell}(1-q^j)}.$$

The Gaussian binomial coefficient is related to the generating function of $p(N, M; n)$ by the equality

$$\sum_{n=0}^{MN} p(N, M, ; n) q^n = \binom{M+N}{M}_q.$$

Rank and Durfee Square

The *rank* of a partition is the largest number k such that the partition contains at least k parts of size at least k. For example, the partition 4 + 3 + 3 + 2 + 1 + 1 has rank 3 because it contains 3 parts that are ≥ 3, but does not contain 4 parts that are ≥ 4. In the Ferrers diagram or Young diagram of a partition of rank r, the $r \times r$ square of entries in the upper-left is known as the Durfee square:

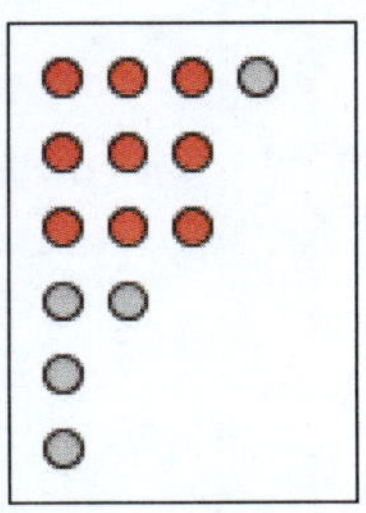

The Durfee square has applications within combinatorics in the proofs of various partition identities. It also has some practical significance in the form of the h-index.

A different statistic is also sometimes called the rank of a partition (or Dyson rank), namely, the difference $\lambda_k - k$ for a partition of k parts with largest part λ_k. This statistic appears in the study of Ramanujan congruences.

Young's Lattice

There is a natural partial order on partitions given by inclusion of Young diagrams. This partially ordered set is known as *Young's lattice.* The lattice was originally defined in the context of representation theory, where it is used to describe the irreducible representations of symmetric groups S_n for all n, together with their branching properties, in characteristic zero. It also has received significant study for its purely combinatorial properties; notably, it is the motivating example of a differential poset.

Pentagonal Number Theorem

In mathematics, the pentagonal number theorem, originally due to Euler, relates the product and series representations of the Euler function. It states that:

$$\prod_{n=1}^{\infty}(1-x^n) = \sum_{k=-\infty}^{\infty}(-1)^k x^{k(3k-1)/2} = 1+\sum_{k=1}^{\infty}(-1)^k\left(x^{k(3k+1)/2}+x^{k(3k+1)/2}+x^{k(3k+1)/2}\right).$$

In other words,

$$(1-x)(1-x^2)(1-x^3)\cdots = 1-x-x^2+x^5+x^7-x^{12}-x^{15}+x^{22}+x^{26}-\cdots.$$

The exponents 1, 2, 5, 7, 12, ... on the right hand side are given by the formula $g_k = k(3k - 1)/2$ for $k = 1, -1, 2, -2, 3, \ldots$ and are called (generalized) pentagonal numbers. This holds as an identity of convergent power series for $|x| < 1$, and also as an identity of formal power series.

A striking feature of this formula is the amount of cancellation in the expansion of the product.

Relation with Partitions

The identity implies a marvelous recurrence for calculating $p(n)$, the number of partitions of n:

$$p(n) = p(n-1)+p(n-2)-p(n-5-p(n-7)+\cdots$$

or more formally,

$$p(n) = \sum_{k}(-1)^{k-1}p(n-gk)$$

where the summation is over all nonzero integers k (positive and negative) and gk is the kth generalized pentagonal number. Since $p(n) = 0$ for all $n < 0$, the series will eventually become zeroes, enabling discrete calculation.

Bijective Proof

The theorem can be interpreted combinatorially in terms of partitions. In particular, the left hand side is a generating function for the number of partitions of n into an even number of distinct parts minus the number of partitions of n into an odd number of distinct parts. Each partition of n into an even number of distinct parts contributes +1 to the coefficient of x^n; each partition into an odd number of distinct parts contributes −1.

For example, the coefficient of x^5 is +1 because there are two ways to split 5 into an even number of distinct parts (4+1 and 3+2), but only one way to do so for an odd number of distinct parts (the one-part partition 5). However, the coefficient of x^{12} is −1 because there are seven ways to partition 12 into an even number of distinct parts, but there are eight ways to partition 12 into an odd number of distinct parts.

This interpretation leads to a proof of the identity via involution (i.e. a bijection which is its own inverse). Consider the Ferrers diagram of any partition of n into distinct parts. For example, the diagram below shows $n = 20$ and the partition $20 = 7 + 6 + 4 + 3$.

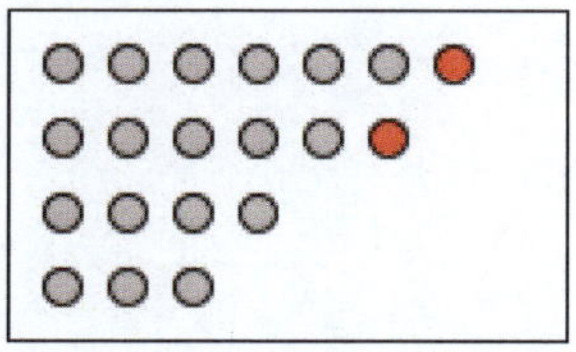

Let m be the number of elements in the smallest row of the diagram ($m = 3$ in the above example). Let s be the number of elements in the rightmost 45 degree line of the diagram ($s = 2$ dots in red above, since $7-1 = 6$, but $6-1 > 4$). If $m > s$, take the rightmost 45-degree line and move it to form a new row, as in the diagram below.

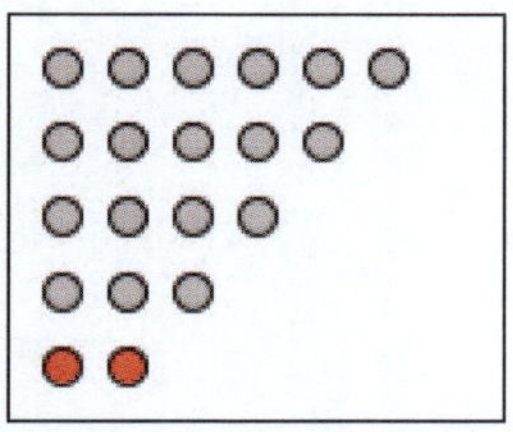

If $m \le s$ (as in our newly formed diagram where $m = 2$, $s = 5$) we may reverse the process by moving the bottom row to form a new 45 degree line (adding 1 element to each of the first m rows), taking us back to the first diagram.

A bit of thought shows that this process always changes the parity of the number of rows, and applying the process twice brings us back to the original diagram. This enables us to pair off Ferrers diagrams contributing 1 and −1 to the x^n term of the series, resulting in a net coefficient of 0. This holds for every term *except* when the process cannot be performed on every Ferrers diagram with n dots. There are two such cases:

1) $m = s$ and the rightmost diagonal and bottom row meet. For example,

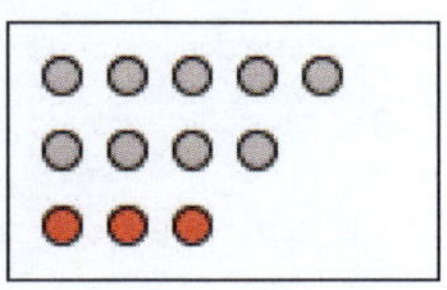

Attempting to perform the operation would lead us to:

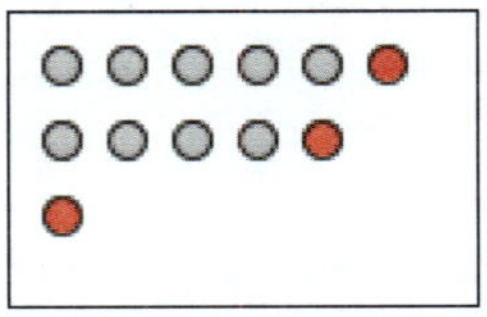

which fails to change the parity of the number of rows, and is not reversible in the sense that performing the operation again does *not* take us back to the original diagram. If there are m elements in the last row of the original diagram, then

$$n = m+(m+1)+(m+2)+\cdots+(2m-1) = \frac{m(3m-1)}{2} = \frac{k(3k-1)}{2}$$

where the new index k is taken to equal m. Note that the sign associated with this partition is $(-1)^s$, which by construction equals $(-1)^m$ and $(-1)^k$.

2) $m = s+1$ and the rightmost diagonal and bottom row meet. For example,

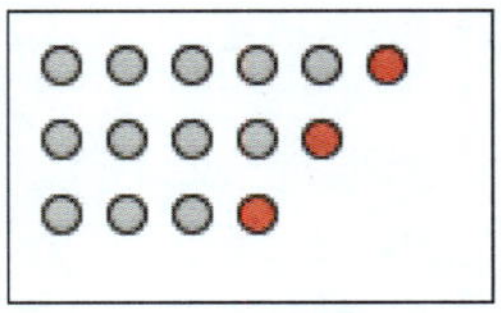

Our operation requires us to move the right diagonal to the bottom row, but that would lead to two rows of three elements, forbidden since we're counting partitions into distinct parts. This is the previous case but with one fewer row, so:

$$n = m+(m+1)+(m+2)+\cdots+(2m-2) = \frac{(m-1)(3m-2)}{2} = \frac{k(3k-1)}{2},$$

where we take $k = 1-m$ (a negative integer). Here the associated sign is $(-1)^s$ with $s = m-1 = -k$, therefore the sign is again $(-1)^k$.

In summary, it has been shown that partitions into an even number of distinct parts and an odd number of distinct parts exactly cancel each other, except if n is a generalized pentagonal number $gk = k(3k-1)/2$, in which case there is exactly one Ferrers diagram left over. But this is precisely what the right side of the identity says should happen.

Partition Recurrence

We can rephrase the above proof, using partitions, which we denote as: $n = \lambda_1 + \lambda_2 + \cdots + \lambda_\ell$, where $\lambda_1 \geq \lambda_2 \geq \ldots \geq \lambda_\ell > 0$. The number of partitions of n is the partition function $p(n)$ having generating function:

$$\sum_{n=0}^{\infty} p(n)x^n = \prod_{k=1}^{\infty}(1-x^k)^{-1}.$$

Note that is the reciprocal of the product on the left hand side of our identity:

$$\left(\sum_{n=0}^{\infty} p(n)x^n\right)\cdot\left(\prod_{n=1}^{\infty}(1-x^n)\right)=1.$$

Let us denote the expansion of our product by $\prod_{n=1}^{\infty}(1-x^n)=\sum_{n=0}^{\infty}a_n x^n$, so that:

$$\left(\sum_{n=0}^{\infty} p(n)x^n\right)\cdot\left(\sum_{n=0}^{\infty}a_n x^n\right)=1.$$

Multiplying out the left hand side and equating coefficients on the two sides, we obtain $a_o\, p(0) =$ 1 and $\sum_{i=0}^{n} p(n-i)a_i = 0$ for all $n \geq 1$. This gives a recurrence relation defining $p(n)$ in terms of a_n, and vice versa a recurrence for a_n in terms of $p(n)$. Thus, our desired result:

$$a_i := \begin{cases} 1 & \text{if } i = \frac{1}{2}(3k^2 \pm k) \text{ and } k \text{ is even} \\ -1 & \text{if } i = \frac{1}{2}(3k^2 \pm k) \text{ and } k \text{ is odd} \\ 0 & \text{otherwise.} \end{cases}$$

for $i \geq 1$ is equivalent to the identity $\sum_i (-1)^i p(n-g_i) = 0$ where $g_i := \frac{1}{2}(3i^2 - i)$ and i ranges over all integers such that $gi \leq n$ (this range includes both positive and negative i, so as to use both kinds of generalized pentagonal numbers). This in turn means:

$$\sum_{i \text{ even}} p(n-g_i) = \sum_{i \text{ odd}} p(n-g).$$

In terms of sets of partitions, this is equivalent to saying that the following sets are of equal cardinality:

$$\mathcal{X} := \bigcup_{i \text{ even}} \mathcal{P}(n-g_i) \quad \text{and} \quad \mathcal{Y} := \bigcup_{i \text{ odd}} \mathcal{P}(n-g_i),$$

where $\mathcal{P}(n)$ denotes the set of all partitions of n. All that remains is to give a bijection from one set to the other, which is accomplished by the function φ from X to Y which maps the partition $\mathcal{P}(n-g_i) \ni \lambda : n-g_i = \lambda_1 + \lambda_2 + \cdots + \lambda_\ell$ to the partition $\lambda' = \varphi(\lambda)$ defined by:

$$\varphi(\lambda) := \begin{cases} \lambda' : n - g_{i-1} = (\ell + 3i - 2) + (\lambda_1 - 1) + \cdots + (\lambda_\ell - 1) \text{ if } \ell + 3i > \lambda_1 \\ \lambda' : n - g_{i+1} = (\lambda_2 + 1) + \cdots + (\lambda_\ell + 1) + \underbrace{1 + \cdots + 1}_{\lambda_1 - \ell - 3i} \quad \text{if } \ell + 3i \leq \lambda_1. \end{cases}$$

This is an involution (a self-inverse mapping), and thus in particular a bijection, which proves our claim and the identity.

Ramanujan's Congruences

In mathematics, Ramanujan's congruences are some remarkable congruences for the partition function $p(n)$. The mathematician Srinivasa Ramanujan discovered the congruences:

$$\begin{aligned} p(5k+4) &\equiv 0 \pmod 5, \\ p(7k+5) &\equiv 0 \pmod 7, \\ p(11k+6) &\equiv 0 \pmod{11}, \end{aligned}$$

This means that:

- If a number is 4 more than a multiple of 5, i.e. it is in the sequence

  4, 9, 14, 19, 24, 29, . . . then the number of its partitions is a multiple of 5.
- If a number is 5 more than a multiple of 7, i.e. it is in the sequence

  5, 12, 19, 26, 33, 40, . . . then the number of its partitions is a multiple of 7.
- If a number is 6 more than a multiple of 11, i.e. it is in the sequence

  6, 17, 28, 39, 50, 61, . . . then the number of its partitions is a multiple of 11.

In his 1919 paper, he proved the first two congruences using the following identities (using q-Pochhammer symbol notation):

$$\sum_{k=0}^{\infty} p(5k+4)q^k = 5\frac{(q^5)_\infty^5}{(q)_\infty^6},$$

$$\sum_{k=0}^{\infty} p(7k+5)q^k = 7\frac{(q^7)_\infty^3}{(q)_\infty^4} + 49q\frac{(q^7)_\infty^7}{(q)_\infty^8}.$$

He then stated that "It appears there are no equally simple properties for any moduli involving primes other than these".

After Ramanujan died in 1920, G. H. Hardy extracted proofs of all three congruences from an unpublished manuscript of Ramanujan on $p(n)$. The proof in this manuscript employs the Eisenstein series.

In 1944, Freeman Dyson defined the rank function and conjectured the existence of a crank function for partitions that would provide a combinatorial proof of Ramanujan's congruences modulo 11. Forty years later, George Andrews and Frank Garvan successfully found such a function, and proved the celebrated result that the crank simultaneously "explains" the three Ramanujan congruences modulo 5, 7 and 11.

In the 1960s, A. O. L. Atkin of the University of Illinois at Chicago discovered additional congruences for small prime moduli. For example:

$$p(11^3 \cdot 13k + 237 \equiv 0 \pmod{13}.$$

Extending the results of A. Atkin, Ken Ono in 2000 proved that there are such Ramanujan congruences modulo every integer coprime to 6. For example, his results give:

$$p(107^4 \cdot 31k + 30064597 \equiv 0 \pmod{31}).$$

Later Ken Ono conjectured that the elusive crank also satisfies exactly the same types of general congruences. This was proved by his Ph.D. student Karl Mahlburg in his 2005 paper Partition Congruences and the Andrews–Garvan–Dyson Crank, linked below. This paper won the first Proceedings of the National Academy of Sciences Paper of the Year prize.

A conceptual explanation for Ramanujan's observation was finally discovered in January 2011 by considering the Hausdorff dimension of the following P function in the l-adic topology:

$$p_\ell(b;z) := \sum_{n=0}^{\infty} p\left(\frac{\ell^b n+1}{24}\right) q^{n/24}.$$

It is seen to have dimension 0 only in the cases where $\ell = 5, 7$ or 11 and since the partition function can be written as a linear combination of these functions this can be considered a formalization and proof of Ramanujan's observation.

In 2001, R.L. Weaver gave an effective algorithm for finding congruences of the partition function, and tabulated 76,065 congruences. This was extended in 2012 by F. Johansson to 22,474,608,014 congruences, one large example being

$$p(999959^4 \cdot 29k + 28995221336976431135321047) \equiv 0 \pmod{29}.$$

MÖBIUS FUNCTION

The classical Möbius function $\mu(n)$ is an important multiplicative function in number theory and combinatorics. The German mathematician August Ferdinand Möbius introduced it in 1832. It is a special case of a more general object in combinatorics.

For any positive integer n, define $\mu(n)$ as the sum of the primitive nth roots of unity. It has values in $\{-1, 0, 1\}$ depending on the factorization of n into prime factors:

- $\mu(n) = 1$ if n is a square-free positive integer with an even number of prime factors.
- $\mu(n) = -1$ if n is a square-free positive integer with an odd number of prime factors.
- $\mu(n) = 0$ if n has a squared prime factor.

The Möbius function can alternatively be represented as:

$$\mu(n) = \delta_{\omega(n)}^{\Omega(n)} \lambda(n),$$

where δ is the Kronecker delta, $\lambda(n)$ is the Liouville function, $\omega(n)$ is the number of distinct prime divisors of n, and $\Omega(n)$ is the number of prime factors of n, counted with multiplicity.

The values of $\mu(n)$ for the first 30 positive numbers are:

n	1	2	3	4	5	6	7	8	9	10
μ(n)	1	−1	−1	0	−1	1	−1	0	0	1

n	11	12	13	14	15	16	17	18	19	20
μ(n)	−1	0	−1	1	1	0	−1	0	−1	0

n	21	22	23	24	25	26	27	28	29	30
μ(n)	1	1	−1	0	0	1	0	0	−1	−1

The first 50 values of the function are plotted below:

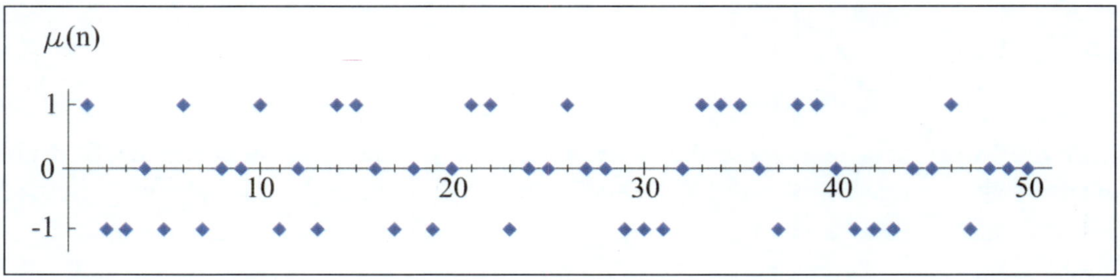

Properties and Applications

Properties

The Möbius function is multiplicative (i.e. $\mu(ab) = \mu(a)\mu(b)$ whenever a and b are coprime).

The sum of the Möbius function over all positive divisors of n (including n itself and 1) is zero except when $n = 1$:

$$\sum_{d|n} \mu(d) = \begin{cases} 1 \text{ if } n = 1, \\ 0 \text{ if } n > 1. \end{cases}$$

The equality above leads to the important Möbius inversion formula and is the main reason why μ is of relevance in the theory of multiplicative and arithmetic functions.

Other applications of $\mu(n)$ in combinatorics are connected with the use of the Pólya enumeration theorem in combinatorial groups and combinatorial enumerations.

There is a formula for calculating the Möbius function without directly knowing the factorization of its argument:

$$\mu(n) = \sum_{\substack{1 \le k \le n \\ \gcd(k,n)=1}} e^{2\pi i \frac{k}{n}},$$

i.e. $\mu(n)$ is the sum of the primitive n-th roots of unity. (However, the computational complexity of this definition is at least the same as that of the Euler product definition).

Proof of the formula for $\sum_{d|n} \mu(d)$

Using:

$$\mu(n) = \sum_{\substack{1 \le k \le n \\ \gcd(k,n)=1}} e^{2\pi i \frac{k}{n}},$$

the formula:

$$\sum_{d|n} \mu(d) = \begin{cases} 1 \text{ if } n = 1, \\ 0 \text{ if } n > 1, \end{cases}$$

can be seen as a consequence of the fact that the nth roots of unity sum to 0, since each nth root of unity is a primitive dth root of unity for exactly one divisor d of n.

However it is also possible to prove this identity from first principles. First note that it is trivially true when $n = 1$. Suppose then that $n > 1$. Then there is a bijection between the factors d of n for which $\mu(d) \neq 0$ and the subsets of the set of all prime factors of n. The asserted result follows from the fact that every non-empty finite set has an equal number of odd- and even-cardinality subsets.

This last fact can be shown easily by induction on the cardinality $|S|$ of a non-empty finite set S. First, if $|S| = 1$, there is exactly one odd-cardinality subset of S, namely S itself, and exactly one even-cardinality subset, namely $\emptyset$. Next, if $|S| > 1$, then divide the subsets of S into two subclasses depending on whether they contain or not some fixed element x in S. There is an obvious bijection between these two subclasses, pairing those subsets that have the same complement relative to the subset $\{x\}$. Also, one of these two subclasses consists of all the subsets of the set $S \setminus \{x\}$, and therefore, by the induction hypothesis, has an equal number of odd- and even-cardinality subsets. These subsets in turn correspond bijectively to the even- and odd-cardinality $\{x\}$-containing subsets of S. The inductive step follows directly from these two bijections.

A related result is that the binomial coefficients exhibit alternating entries of odd and even power which sum symmetrically.

Mertens Function

In number theory another arithmetic function closely related to the Möbius function is the Mertens function, defined by:

$$M(n) = \sum_{k=1}^{n} \mu(k),$$

for every natural number n. This function is closely linked with the positions of zeroes of the Riemann zeta function.

From the formula:

$$\mu(n) = \sum_{\substack{1 \le k \le n \\ \gcd(k,n)=1}} e^{2\pi i \frac{k}{n}},$$

it follows that the Mertens function is given by:

$$M(n) = -1 + \sum_{a \in F_n} e^{2\pi i a},$$

where F_n is the Farey sequence of order n.

This formula is used in the proof of the Franel–Landau theorem.

Applications

Mathematical Series

The Dirichlet series that generates the Möbius function is the (multiplicative) inverse of the Riemann zeta function; if s is a complex number with real part larger than 1 we have:

$$\sum_{n=1}^{\infty} \frac{\mu(n)}{n^s} = \frac{1}{\zeta(s)}.$$

This may be seen from its Euler product:

$$\frac{1}{\zeta(s)} = \prod_{p \in \mathbb{P}} \left(1 - \frac{1}{p^s}\right) = \left(1 - \frac{1}{2^s}\right)\left(1 - \frac{1}{3^s}\right)\left(1 - \frac{1}{5^s}\right)\cdots.$$

The Lambert series for the Möbius function is:

$$\sum_{n=1}^{\infty} \frac{\mu(n) q^n}{1 - q^n} = q,$$

which converges for $|q| < 1$. For prime $\alpha \geq 2$, we also have:

$$\sum_{n=1}^{\infty} \frac{\mu(\alpha n) q^n}{1 - q^n} = \sum_{n \geq} q^{\alpha^n}, |q| < 1.$$

The Möbius function is a sum of the number of integer points on n-dimensional hyperboloids:

$$\mu(n)_{n>1} = -\sum_{\substack{a \geq 2 \\ a=n}} 1 + \sum_{\substack{a \geq 2 \\ ab=n}} \sum_{b \geq 2} 1 - \sum_{a \geq 2} \sum_{\substack{b \geq 2 \\ abc=n}} \sum_{c \geq 2} 1 + \sum_{a \geq 2} \sum_{b \geq 2} \sum_{\substack{c \geq 2 \\ abcd=n}} \sum_{d \geq 2} 1 - \cdots.$$

Algebraic Number Theory

Gauss proved that for a prime number p the sum of its primitive roots is congruent to $\mu(p - 1) \pmod{p}$.

If F_q denotes the finite field of order q (where q is necessarily a prime power), then the number N of monic irreducible polynomials of degree n over F_q is given by:

$$N(q,n) = \frac{1}{n} \sum_{d|n} \mu(d) q^{n/d}.$$

Average Order

The average order of the Möbius function is zero. This statement is, in fact, equivalent to the prime number theorem.

$\mu(n)$ Sections

$\mu(n) = 0$ if and only if n is divisible by the square of a prime. The first numbers with this property are: 4, 8, 9, 12, 16, 18, 20, 24, 25, 27, 28, 32, 36, 40, 44, 45, 48, 49, 50, 52, 54, 56, 60, 63, ...

If n is prime, then $\mu(n) = -1$, but the converse is not true. The first non-prime n for which $\mu(n) = -1$ is $30 = 2 \times 3 \times 5$. The first such numbers with three distinct prime factors (sphenic numbers) are: 30, 42, 66, 70, 78, 102, 105, 110, 114, 130, 138, 154, 165, 170, 174, 182, 186, 190, 195, 222, ... and the first such numbers with 5 distinct prime factors are: 2310, 2730, 3570, 3990, 4290, 4830, 5610, 6006, 6090, 6270, 6510, 6630, 7410, 7590, 7770, 7854, 8610, 8778, 8970, 9030, 9282, 9570, 9690, ...

Generalizations

Incidence Algebras

In combinatorics, every locally finite partially ordered set (poset) is assigned an incidence algebra. One distinguished member of this algebra is that poset's "Möbius function". The classical Möbius function treated in this topic is essentially equal to the Möbius function of the set of all positive integers partially ordered by divisibility.

Popovici's Function

Constantin Popovici defined a generalised Möbius function $\mu_k = \mu * \ldots * \mu$ to be the k-fold Dirichlet convolution of the Möbius function with itself. It is thus again a multiplicative function with:

$$\mu_k(p^a) = (-1)^a \binom{k}{a},$$

where the binomial coefficient is taken to be zero if $a > k$. The definition may be extended to complex k by reading the binomial as a polynomial in k.

Physics

The Möbius function also arises in the primon gas or free Riemann gas model of supersymmetry. In this theory, the fundamental particles or "primons" have energies $\log p$. Under second quantization, multiparticle excitations are considered; these are given by $\log n$ for any natural number n. This follows from the fact that the factorization of the natural numbers into primes is unique.

In the free Riemann gas, any natural number can occur, if the primons are taken as bosons. If they are taken as fermions, then the Pauli exclusion principle excludes squares. The operator $(-1)^F$ that distinguishes fermions and bosons is then none other than the Möbius function $\mu(n)$.

The free Riemann gas has a number of other interesting connections to number theory, including the fact that the partition function is the Riemann zeta function. This idea underlies Alain Connes's attempted proof of the Riemann hypothesis.

ERDŐS–KAC THEOREM

In number theory, the Erdős–Kac theorem, named after Paul Erdős and Mark Kac, and also known as the fundamental theorem of probabilistic number theory, states that if $\omega(n)$ is the number of distinct prime factors of n, then, loosely speaking, the probability distribution of:

$$\frac{\omega(n)-\log\log n}{\sqrt{\log\log n}},$$

is the standard normal distribution. This is an extension of the Hardy–Ramanujan theorem, which states that the normal order of $\omega(n)$ is $\log \log n$ with a typical error of size $\sqrt{\log\log n}$.

Precise Statement

For any fixed $a < b$,

$$\lim_{x\to\infty}\left(\frac{1}{x}\cdot\#\left\{n\le x : a\le\frac{\omega(n)-\log\log n}{\sqrt{\log\log n}}\le b\right\}\right)=\Phi(a,b).$$

where $\Phi(a,b)$ is the normal (or "Gaussian") distribution, defined as:

$$\Phi(a,b)=\frac{1}{\sqrt{2\pi}}\int_a^b e^{-t^2/2dt}.$$

More generally, if f(n) is a strongly additive function $(f(p_1 \ldots p_k)=f(p_1)+\cdots+f(p_k))$ with $|f(p)|\le 1$ for all prime p, then:

$$\lim_{x\to\infty}\left(\frac{1}{x}\cdot\#\left\{n\le x : a\le\frac{f(n)-A(n)}{B(n)}\le b\right\}\right)=\Phi(a,b),$$

with,

$$A(n)=\sum_{p\le n}\frac{f(p)}{p},\quad B(n)=\sqrt{\sum_{p\le n}\frac{f(p)^2}{p}}.$$

Kac's Original Heuristic

Intuitively, Kac's heuristic for the result says that if n is a randomly chosen large integer, then the number of distinct prime factors of n is approximately normally distributed with mean and variance log log n. This comes from the fact that given a random natural number n, the events "the number n is divisible by some prime p" for each p are mutually independent.

Now, denoting the event "the number n is divisible by p" by n_p, consider the following sum of indicator random variables:

$$I_{n_2} + I_{n_3} + I_{n_5} + I_{n_7} + \cdots$$

This sum counts how many distinct prime factors our random natural number n has. It can be shown that this sum satisfies the Lindeberg condition, and therefore the Lindeberg central limit theorem guarantees that after appropriate rescaling, the above expression will be Gaussian.

The actual proof of the theorem, due to Erdős, uses sieve theory to make rigorous the above intuition.

Numerical examples:

The Erdős–Kac theorem means that the construction of a number around one billion requires on average three primes.

For example, 1,000,000,003 = 23 × 307 × 141623. The following table provides a numerical summary of the growth of the average number of distinct prime factors of a natural number n with increasing n.

n	Number of digits in n	Average number of distinct primes	Standard deviation
1,000	4	2	1.4
1,000,000,000	10	3	1.7
1,000,000,000,000,000,000,000,000	25	4	2
10^{65}	66	5	2.2
$10^{9,566}$	9,567	10	3.2
$10^{210,704,568}$	210,704,569	20	4.5
$10^{10^{22}}$	10^{22}+1	50	7.1
$10^{10^{44}}$	10^{44}+1	100	10
$10^{10^{434}}$	10^{434}+1	1000	31.6

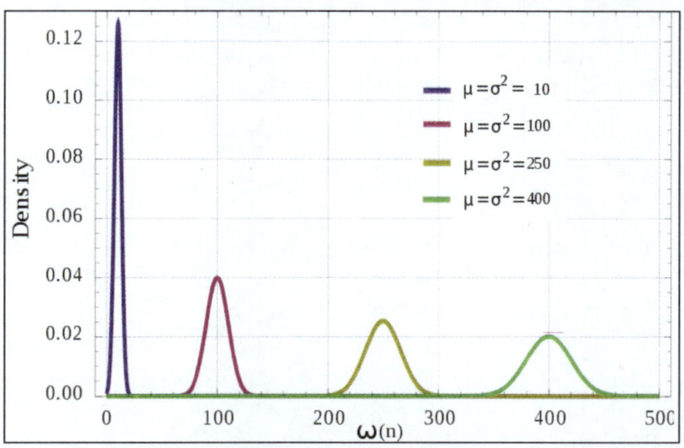

A spreading Gaussian distribution of distinct primes illustrating the Erdos-Kac theorem.

Around 12.6% of 10,000 digit numbers are constructed from 10 distinct prime numbers and around 68% are constructed from between 7 and 13 primes.

A hollow sphere the size of the planet Earth filled with fine sand would have around 10^{33} grains. A volume the size of the observable universe would have around 10^{93} grains of sand. There might be room for 10^{185} quantum strings in such a universe.

Numbers of this magnitude—with 186 digits—would require on average only 6 primes for construction.

It is very difficult if not impossible to discover this result empirically, as the Gaussian only shows up when n starts getting to be around 10^{100}. More precisely, Rényi and Turán showed that the best possible uniform asymptotic bound on the error in the approximation to a Gaussian is $O\left(\frac{1}{\sqrt{\log\log n}}\right)$.

BELL SERIES

In mathematics, the Bell series is a formal power series used to study properties of arithmetical functions. Bell series were introduced and developed by Eric Temple Bell.

Given an arithmetic function f and a prime p, define the formal power series $f_p(x)$, called the Bell series of f modulo p as:

$$f_p(x) = \sum_{n=0}^{\infty} f(p^n)x^n.$$

Two multiplicative functions can be shown to be identical if all of their Bell series are equal; this is sometimes called the uniqueness theorem: given multiplicative functions f and g, one has $f = g$ if and only if:

$f_p(x) = g_p(x)$ for all primes p.

Two series may be multiplied (sometimes called the *multiplication theorem*): For any two arithmetic functions f and g, let $h = f * g$ be their Dirichlet convolution. Then for every prime p, one has:

$$h_p(x) = f_p(x)g_p(x).$$

In particular, this makes it trivial to find the Bell series of a Dirichlet inverse.

If f is completely multiplicative, then formally:

$$f_p(x) = \frac{1}{1 - f(p)x}.$$

Examples:

The following is a table of the Bell series of well-known arithmetic functions:

- The Möbius function μ has $\mu_p(x) = 1 - x$.
- The Mobius function squared has $\mu_p^2(x) = 1 + x$.
- Euler's totient φ has $\varphi_p(x) = \dfrac{1-x}{1-px}$.
- The multiplicative identity of the Dirichlet convolution δ has $\delta_p(x) = 1$.
- The Liouville function λ has $\lambda_p(x) = \dfrac{1}{1+x}$.
- The power function Id_k has $(\mathrm{Id}_k)_p(x) = \dfrac{1}{1-p^k x}$. Here, Id_k is the completely multiplicative function $\mathrm{Id}_k(n) = n^k$.
- The divisor function σ_k has $(\sigma_k)_p(x) = \dfrac{1}{1-(1+p^k)x + p^k x^2}$.
- The unit function satisfies $1_p(x) = (1-x)^{-1}$, i.e., is the geometric series.
- If $f(n) = 2^{\omega(n)} = \sum_{d|n} \mu^2$ is the power of the prime omega function, then $f_p(x) = \dfrac{1+x}{1-x}$.
- Suppose that f is multiplicative and g is any arithmetic function satisfying $f(p^{n+1}) = f(p)f(p^n) - g(p)f(p^{n-1})$ for all primes p and $n \geq 1$. Then $f_p(x) = \left(1 - f(p)x + \right.$ $\left. + g(p)x^2\right)^{-1}$.
- If $\mu_k(n) = \sum_{d^k|n} \mu_{k-1}\left(\dfrac{n}{d^k}\right)\mu_{k-1}\left(\dfrac{n}{d}\right)$ denotes the Mobius function of order k, then $(\mu_k)_p(x) = \dfrac{1-2x^k + x^{k+1}}{1-x}$.

LAMBERT SERIES

A Lambert series is a series of the form:

$$F(x) \equiv \sum_{n=1}^{\infty} a_n \frac{x^n}{1-x^n} \quad 1$$

for $|x| < 1$. Then:

$$F(x) = \sum_{n=1}^{\infty} a_n \sum_{m=1}^{\infty} x^{mn}$$

$$= \sum_{N=1}^{\infty} b_N x^N,$$

where,

$$b_N \sum_{n/N} a_n.$$

The particular case $a_n = 1$ is sometimes denoted:

$$L(\beta) = \sum_{n=1}^{\infty} \frac{\beta^n}{1-\beta^n}$$

$$= \sum_{n=1}^{\infty} \frac{1}{\beta^{-n} - 1}$$

$$= \frac{\psi_\beta(1) + \text{In}(1-\beta)}{\text{In}\,\beta}$$

For $|\beta| < 1$ where $\psi_q(z)$ is a q-polygamma function. Special cases and related sums include:

$$\sum_{n=1}^{\infty} \frac{\beta^n}{1+\beta^{2n}} = \frac{1}{4}\left[\vartheta_3^2(\beta) - 1\right]$$

$$\sum_{n=1}^{\infty} \frac{\beta^{2n+1}}{1+\beta^{4n+2}} = \frac{1}{4}\left[\vartheta_3^2(\beta) - \varphi_2^2(\beta)\right]$$

$$= \frac{1}{4}\vartheta_2^2(\beta^2)$$

$$\sum_{n=1}^{\infty} \frac{\beta^n}{1-\beta^{2n}} = L(\beta) - L(\beta^2)$$

$$\sum_{n=1}^{\infty} \frac{\beta^{2n+1}}{1-\beta^{4n+2}} = L(\beta) - 2L(\beta^2) + L(\beta^4).$$

which arise in the reciprocal Fibonacci and reciprocal Lucas constants. Some beautiful series of this type include:

$$\sum_{n=1}^{\infty} \frac{\mu(n)x^n}{1-x^n} = x$$

$$\sum_{n=1}^{\infty} \frac{\phi(n)x^n}{1-x^n} = \frac{x}{(1-x)^2}$$

$$\sum_{n=1}^{\infty} \frac{x^n}{1-x^n} = \sum_{n=1}^{\infty} d(n)x^n$$

$$= \frac{\psi_x(1) + \text{In}(1-x)}{\text{In}\,x}$$

$$\sum_{n=1}^{\infty}\frac{n^k x^n}{1-x^n}=\sum_{n=1}^{\infty}\sigma_k(n)x^n$$

$$4\sum_{n=1}^{\infty}\frac{(-1)^{n+1}x^{2n+1}}{1-x^{2n+1}}=\sum_{n=1}^{\infty}r_2(n)x^2$$

$$=\vartheta_3^2(x)-1$$

$$\sum_{n=1}^{\infty}\frac{\lambda(n)x^n}{1-x^n}=\sum_{n=1}^{\infty}x^{n^2}$$

$$=\frac{1}{2}[\vartheta_3(x)-1]$$

$$\sum_{n=1}^{\infty}\frac{\text{Isb}(n)x^n}{1-x^n}=\frac{\text{In}(1-x^2)+\psi_{x^2}\left(\frac{1}{2}\right)}{\text{In}(x^2)},$$

Where $\mu(n)$ is the Möbius function, $\phi(n)$ is the totient function, $d(n)=\sigma_0(n)$ is the number of divisors of $n,\psi_q(z)$ is the q-polygamma function, $\sigma_k(n)$ is the divisor function, $r(n)$ is the number of representations of n in the form $n=A^2+B^2$ where A and B are rational integers $\vartheta_3(q)$ is a Jacobi elliptic function $\lambda(n)$ is the Liouville function, and $\text{Isb}(n)$ is the least significant bit of n.

References

- Arithmetic-function, science: britannica.com, Retrieved 31 March, 2019
- Humphries, peter (2013). "the distribution of weighted sums of the liouville function and pólya's conjecture". Journal of number theory. 133 (2): 545–582. Arxiv:1108.1524. Doi:10.1016/j.jnt.2012.08.011
- Divisorfunction: wolfram.com, Retrieved 14 July, 2019
- Hugh l. Montgomery; robert c. Vaughan (2007). Multiplicative number theory i. Classical theory. Cambridge tracts in advanced mathematics. 97. Cambridge: cambridge univ. Press. P. 38. Isbn 0-521-84903-9
- Ewell, john a. (2004), "recurrences for the partition function and its relatives", the rocky mountain journal of mathematics, 34 (2): 619–627, doi:10.1216/rmjm/1181069871, jstor 44238988, mr 2072798
- Lambertseries: wolfram.com, Retrieved 17 May, 2019

CHAPTER 3

Analytic Number Theory

Analytic number theory uses various methods from mathematical analysis for solving problems related to integers. Prime number theorem, totient function, prime triplet, prime quadruplet, Waring's problem, Sierpinski number, etc. are some of its concepts. This chapter has been carefully written to provide an extensive understanding of the subject.

Analytic number theory is the branch of number theory which uses real and complex analysis to investigate various properties of integers and prime numbers. Examples of topics falling under analytic number theory include Dirichlet L-series, the Riemann zeta function $\zeta(s)$ the totient function $\phi(n)$ and the prime number theorem.

PRIME NUMBER

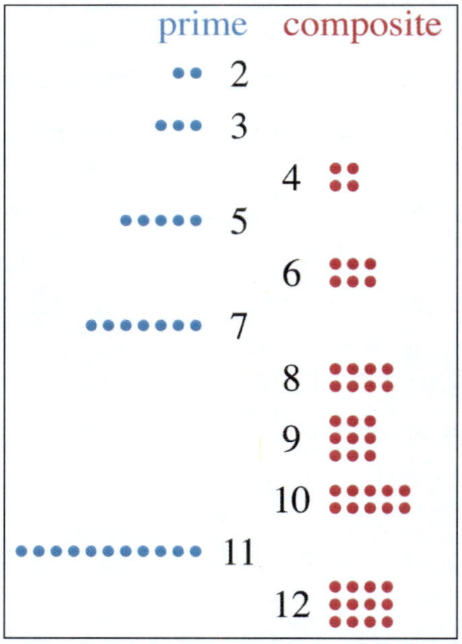

The prime numbers are the natural numbers greater than one that are not products of two smaller natural numbers.

A prime number (or a prime) is a natural number greater than 1 that cannot be formed by multiplying two smaller natural numbers. A natural number greater than 1 that is not prime is called a composite number. For example, 5 is prime because the only ways of writing it as a product, 1 × 5 or 5 × 1, involve 5 itself. However, 6 is composite because it is the product of two numbers (2 × 3) that are both smaller than 6. Primes are central in number theory because of the fundamental theorem of arithmetic: every natural number greater than 1 is either a prime itself or can be factorized as a product of primes that is unique up to their order.

The property of being prime is called primality. A simple but slow method of checking the primality of a given number n, called trial division, tests whether n is a multiple of any integer between 2 and $\sqrt{n}$. Faster algorithms include the Miller–Rabin primality test, which is fast but has a small chance of error, and the AKS primality test, which always produces the correct answer in polynomial time but is too slow to be practical. Particularly fast methods are available for numbers of special forms, such as Mersenne numbers. As of December 2018 the largest known prime number has 24,862,048 decimal digits.

There are infinitely many primes, as demonstrated by Euclid around 300 BC. No known simple formula separates prime numbers from composite numbers. However, the distribution of primes within the natural numbers in the large can be statistically modelled. The first result in that direction is the prime number theorem, proven at the end of the 19th century, which says that the probability of a randomly chosen number being prime is inversely proportional to its number of digits, that is, to its logarithm.

Several historical questions regarding prime numbers are still unsolved. These include Goldbach's conjecture, that every even integer greater than 2 can be expressed as the sum of two primes, and the twin prime conjecture, that there are infinitely many pairs of primes having just one even number between them. Such questions spurred the development of various branches of number theory, focusing on analytic or algebraic aspects of numbers. Primes are used in several routines in information technology, such as public-key cryptography, which relies on the difficulty of factoring large numbers into their prime factors. In abstract algebra, objects that behave in a generalized way like prime numbers include prime elements and prime ideals.

A natural number (1, 2, 3, 4, 5, 6, etc.) is called a *prime number* (or a *prime*) if it is greater than 1 and cannot be written as the product of two smaller natural numbers. The numbers greater than 1 that are not prime are called composite numbers. In other words, n is prime if n items cannot be divided up into smaller equal-size groups of more than one item, or if it is not possible to arrange n dots into a rectangular grid that is more than one dot wide and more than one dot high. For example, among the numbers 1 through 6, the numbers 2, 3, and 5 are the prime numbers, as there are no other numbers that divide them evenly (without a remainder). 1 is not prime, as it is specifically excluded in the definition. 4 = 2 × 2 and 6 = 2 × 3 are both composite.

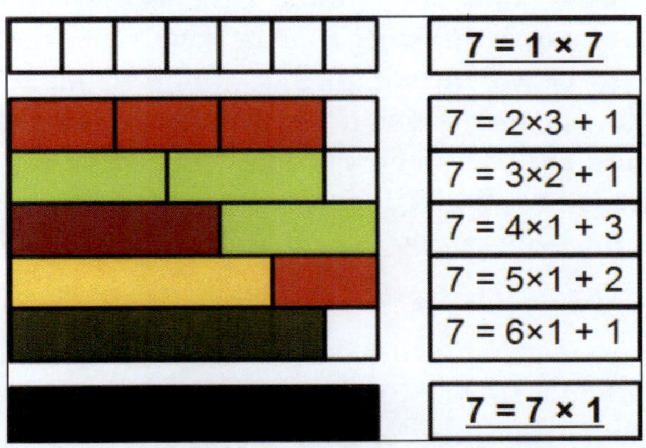

Demonstration, with Cuisenaire rods, that 7 is prime, because none of 2, 3, 4, 5, or 6 divide it evenly.

The divisors of a natural number n are the natural numbers that divide n evenly. Every natural number has both 1 and itself as a divisor. If it has any other divisor, it cannot be prime. This idea leads to a different but equivalent definition of the primes: they are the numbers with exactly two positive divisors, 1 and the number itself. Yet another way to express the same thing is that a number n is prime if it is greater than one and if none of the numbers $2,3,\ldots,n-1$ divides n evenly.

The first 25 prime numbers (all the prime numbers less than 100) are: 2, 3, 5, 7, 11, 13, 17, 19, 23, 29, 31, 37, 41, 43, 47, 53, 59, 61, 67, 71, 73, 79, 83, 89, 97.

No even number n greater than 2 is prime because any such number can be expressed as the product $2 \times n/2$. Therefore, every prime number other than 2 is an odd number, and is called an *odd prime*. Similarly, when written in the usual decimal system, all prime numbers larger than 5 end in 1, 3, 7, or 9. The numbers that end with other digits are all composite: decimal numbers that end in 0, 2, 4, 6, or 8 are even, and decimal numbers that end in 0 or 5 are divisible by 5.

The set of all primes is sometimes denoted by $\mathbf{P}$ (a boldface capital *P*) or by $\mathbb{P}$ (a blackboard bold capital P).

Elementary Properties

Unique Factorization

Writing a number as a product of prime numbers is called a *prime factorization* of the number. For example:

$$\begin{aligned} 34866 &= 2 \cdot 3 \cdot 3 \cdot 13 \cdot 149 \\ &= 2 \cdot 3^2 \cdot 13 \cdot 149. \end{aligned}$$

The terms in the product are called *prime factors*. The same prime factor may occur more than once; this example has two copies of the prime factor 3. When a prime occurs multiple times, exponentiation can be used to group together multiple copies of the same prime number: for example, in the second way of writing the product above, 3^2 denotes the square or second power of 3.

The central importance of prime numbers to number theory and mathematics in general stems from the *fundamental theorem of arithmetic*. This theorem states that every integer larger than 1 can be written as a product of one or more primes. More strongly, this product is unique in the sense that any two prime factorizations of the same number will have the same numbers of copies of the same primes, although their ordering may differ. So, although there are many different ways of finding a factorization using an integer factorization algorithm, they all must produce the same result. Primes can thus be considered the "basic building blocks" of the natural numbers.

Some proofs of the uniqueness of prime factorizations are based on Euclid's lemma: If p is a prime number and p divides a product ab of integers a and b, then p divides a or divides p (or both). Conversely, if a number p has the property that when it divides a product it always divides at least one factor of the product, then p must be prime.

Infiniteness

There are infinitely many prime numbers. Another way of saying this is that the sequence

2, 3, 5, 7, 11, 13, ...

of prime numbers never ends. This statement is referred to as *Euclid's theorem* in honor of the ancient Greek mathematician Euclid, since the first known proof for this statement is attributed to him. Many more proofs of the infinitude of primes are known, including an analytical proof by Euler, Goldbach's proof based on Fermat numbers, Furstenberg's proof using general topology, and Kummer's elegant proof.

Euclid's proof shows that every finite list of primes is incomplete. The key idea is to multiply together the primes in any given list and add 1. If the list consists of the primes $p_1, p_2, \ldots, p_n$, this gives the number:

$$N = 1 + p_1 \cdot p_2 \cdots p_n.$$

By the fundamental theorem, N has a prime factorization:

$$N = p_1' \cdot p_2' \cdots p_m'$$

with one or more prime factors. N is evenly divisible by each of these factors, but N has a remainder of one when divided by any of the prime numbers in the given list, so none of the prime factors of N can be in the given list. Because there is no finite list of all the primes, there must be infinitely many primes.

The numbers formed by adding one to the products of the smallest primes are called Euclid numbers. The first five of them are prime, but the sixth,

$$1 + (2 \cdot 3 \cdot 5 \cdot 7 \cdot 11 \cdot 13) = 30031 = 59 \cdot 509,$$

is a composite number.

Formulas for Primes

There is no known efficient formula for primes. For example, there is no non-constant polynomial, even in several variables, that takes *only* prime values. However, there are numerous expressions that do encode all primes, or only primes. One possible formula is based on Wilson's theorem and generates the number 2 many times and all other primes exactly once. There is also a set of Diophantine equations in nine variables and one parameter with the following property: the parameter is prime if and only if the resulting system of equations has a solution over the natural numbers. This can be used to obtain a single formula with the property that all its *positive* values are prime.

Other examples of prime-generating formulas come from Mills' theorem and a theorem of Wright. These assert that there are real constants $A > 1$ and μ such that

$$\left\lfloor A^{3^n} \right\rfloor \text{ and } \left\lfloor 2^{\cdot^{\cdot^{2^{2^{\mu}}}}} \right\rfloor$$

are prime for any natural number n in the first formula, and any number of exponents in the second formula. Here $\lfloor \cdot \rfloor$ represents the floor function, the largest integer less than or equal to the number in question. However, these are not useful for generating primes, as the primes must be generated first in order to compute the values of A or μ.

Open Questions

Many conjectures revolving about primes have been posed. Often having an elementary formulation, many of these conjectures have withstood proof for decades: all four of Landau's problems from 1912 are still unsolved. One of them is Goldbach's conjecture, which asserts that every even integer n greater than 2 can be written as a sum of two primes. As of 2014, this conjecture has been verified for all numbers up to $n = 4{\cdot}10^{18}$. Weaker statements than this have been proven, for example, Vinogradov's theorem says that every sufficiently large odd integer can be written as a sum of three primes. Chen's theorem says that every sufficiently large even number can be expressed as the sum of a prime and a semiprime (the product of two primes). Also, any even integer greater than 10 can be written as the sum of six primes. The branch of number theory studying such questions is called additive number theory.

Another type of problem concerns prime gaps, the differences between consecutive primes. The existence of arbitrarily large prime gaps can be seen by noting that the sequence $n!+2, n!+3, \ldots, n!+n$ consists of $n-1$ composite numbers, for any natural number n. However, large prime gaps occur much earlier than this argument shows. For example, the first prime gap of length 8 is between the primes 89 and 97, much smaller than $8! = 40320$. It is conjectured that there are infinitely many twin primes, pairs of primes with difference 2; this is the twin prime conjecture. Polignac's conjecture states more generally that for every positive integer k, there are infinitely many pairs of consecutive primes that differ by $2k$. Andrica's conjecture, Brocard's conjecture, Legendre's conjecture, and Oppermann's conjecture all suggest that the largest gaps between primes from 1 to n should be at most approximately $\sqrt{n}$, a result that is known to follow from the Riemann hypothesis, while the much stronger Cramér conjecture sets the largest gap size at $O((\log n)^2)$. Prime gaps can be generalized to prime k-tuples, patterns in the differences between more than two prime numbers. Their infinitude and density are the subject of the first Hardy–Littlewood conjecture, which can be motivated by the heuristic that the prime numbers behave similarly to a random sequence of numbers with density given by the prime number theorem.

Analytic Properties

Analytic number theory studies number theory through the lens of continuous functions, limits, infinite series, and the related mathematics of the infinite and infinitesimal.

This area of study began with Leonhard Euler and his first major result, the solution to the Basel problem. The problem asked for the value of the infinite sum $1+\frac{1}{4}+\frac{1}{9}+\frac{1}{16}+\ldots$, which today can be recognized as the value $\zeta(2)$ of the Riemann zeta function. This function is closely connected to the prime numbers and to one of the most significant unsolved problems in mathematics, the Riemann hypothesis. Euler showed that $\zeta(2) = \pi^2/6$. The reciprocal of this number, $6/\pi^2$, is the limiting probability that two random numbers selected uniformly from a large range are relatively prime (have no factors in common).

The distribution of primes in the large, such as the question how many primes are smaller than a given, large threshold, is described by the prime number theorem, but no efficient formula for the n-th prime is known. Dirichlet's theorem on arithmetic progressions, in its basic form, asserts that linear polynomials,

$$p(n) = a + bn$$

with relatively prime integers a and b take infinitely many prime values. Stronger forms of the theorem state that the sum of the reciprocals of these prime values diverges, and that different linear polynomials with the same b have approximately the same proportions of primes. Although conjectures have been formulated about the proportions of primes in higher-degree polynomials, they remain unproven, and it is unknown whether there exists a quadratic polynomial that (for integer arguments) is prime infinitely often.

Analytical Proof of Euclid's Theorem

Euler's proof that there are infinitely many primes considers the sums of reciprocals of primes,

$$\frac{1}{2} + \frac{1}{3} + \frac{1}{5} + \frac{1}{7} + \cdots + \frac{1}{p}.$$

Euler showed that, for any arbitrary real number x, there exists a prime p for which this sum is bigger than x. This shows that there are infinitely many primes, because if there were finitely many primes the sum would reach its maximum value at the biggest prime rather than growing past every x. The growth rate of this sum is described more precisely by Mertens' second theorem. For comparison, the sum

$$\frac{1}{1^2} + \frac{1}{2^2} + \frac{1}{3^2} + \cdots + \frac{1}{n^2},$$

does not grow to infinity as n goes to infinity. In this sense, prime numbers occur more often than squares of natural numbers, although both sets are infinite. Brun's theorem states that the sum of the reciprocals of twin primes,

$$\left(\frac{1}{3} + \frac{1}{5}\right) + \left(\frac{1}{5} + \frac{1}{7}\right) + \left(\frac{1}{11} + \frac{1}{13}\right) + \cdots,$$

is finite. Because of Brun's theorem, it is not possible to use Euler's method to solve the twin prime conjecture, that there exist infinitely many twin primes.

Number of Primes Below a Given Bound

The relative error of $\frac{n}{\log n}$ and the logarithmic integral $Li(n)$ as approximations to the prime-counting function. Both relative errors decrease to zero as n grows, but the convergence to zero is much more rapid for the logarithmic integral.

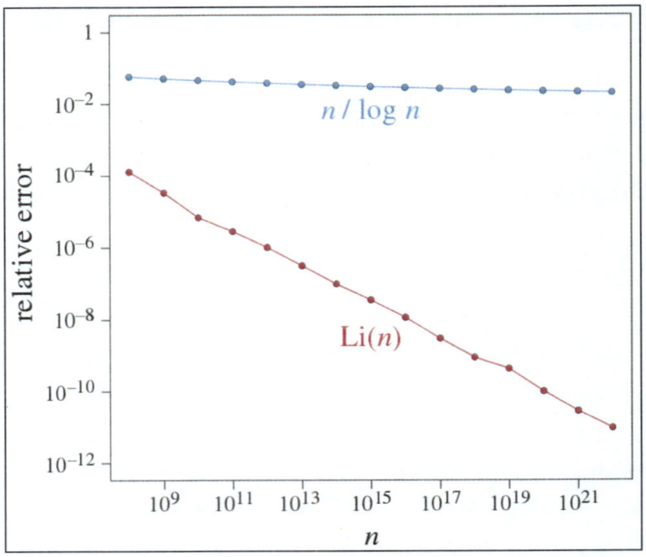

The prime counting function $\pi(n)$ is defined as the number of primes not greater than n. For example, $\pi(11) = 5$, since there are five primes less than or equal to 11. Methods such as the Meissel–Lehmer algorithm can compute exact values of $\pi(n)$ faster than it would be possible to list each prime up to n. The prime number theorem states that $\pi(n)$ is asymptotic to $n / \log n$, which is denoted as:

$$\pi(n) \sim \frac{n}{\log n},$$

and means that the ratio of $\pi(n)$ to the right-hand fraction approaches 1 as n grows to infinity. This implies that the likelihood that a randomly chosen numbers less than n is prime is (approximately) inversely proportional to the number of digits in n. It also implies that the nth prime number is proportional to $n / \log n$ and therefore that the average size of a prime gap is proportional to $\log n$. A more accurate estimate for $\pi(n)$ is given by the offset logarithmic integral:

$$\pi(n) \sim \mathrm{Li}(n) = \int_2^n \frac{dt}{\log t}.$$

Arithmetic Progressions

An arithmetic progression is a finite or infinite sequence of numbers such that consecutive numbers in the sequence all have the same difference. This difference is called the modulus of the progression. For example, 3, 12, 21, 30, 39, ...,is an infinite arithmetic progression with modulus 9. In an arithmetic progression, all the numbers have the same remainder when divided by the modulus; in this example, the remainder is 3. Because both the modulus 9 and the remainder 3 are multiples of 3, so is every element in the sequence. Therefore, this progression contains only one prime number, 3 itself. In general, the infinite progression $a, a+q, a+2q, a+3q, \ldots$ can have more than one prime only when its remainder a and modulus q are relatively prime. If they are relatively prime, Dirichlet's theorem on arithmetic progressions asserts that the progression contains infinitely many primes.

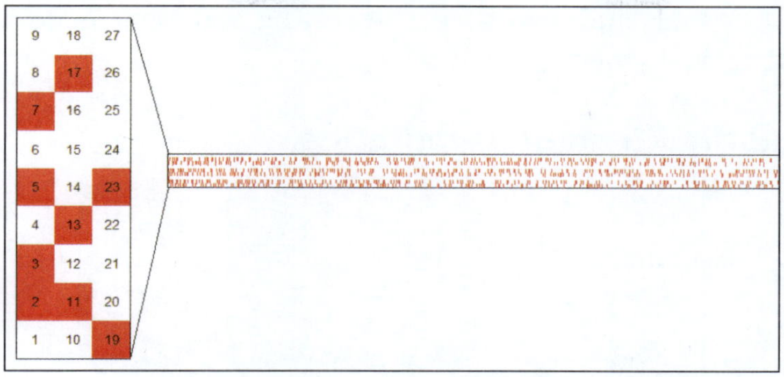

Primes in the arithmetic progressions modulo 9. Each row of the thin horizontal band shows one of the nine possible progressions mod 9, with prime numbers marked in red. The progressions of numbers that are 0, 3, or 6 mod 9 contain at most one prime number (the number 3); the remaining progressions of numbers that are 2, 4, 5, 7, and 8 mod 9 have infinitely many prime numbers, with similar numbers of primes in each progression.

The Green–Tao theorem shows that there are arbitrarily long finite arithmetic progressions consisting only of primes.

Prime Values of Quadratic Polynomials

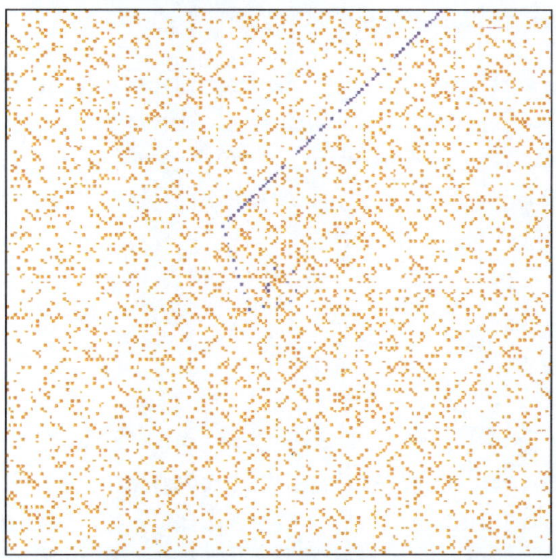

The Ulam spiral. Prime numbers (red) cluster on some diagonals and not others. Prime values of $4n^2 - 2n + 41$ are shown in blue.

Euler noted that the function $n^2 - n + 41$ yields prime numbers for $1 \leq n \leq 40$, although composite numbers appear among its later values. The search for an explanation for this phenomenon led to the deep algebraic number theory of Heegner numbers and the class number problem. The Hardy-Littlewood conjecture F predicts the density of primes among the values of quadratic polynomials with integer coefficients in terms of the logarithmic integral and the polynomial coefficients. No quadratic polynomial has been proven to take infinitely many prime values.

The Ulam spiral arranges the natural numbers in a two-dimensional grid, spiraling in concentric squares surrounding the origin with the prime numbers highlighted. Visually, the primes appear to cluster on certain diagonals and not others, suggesting that some quadratic polynomials take prime values more often than others.

Zeta Function and the Riemann Hypothesis

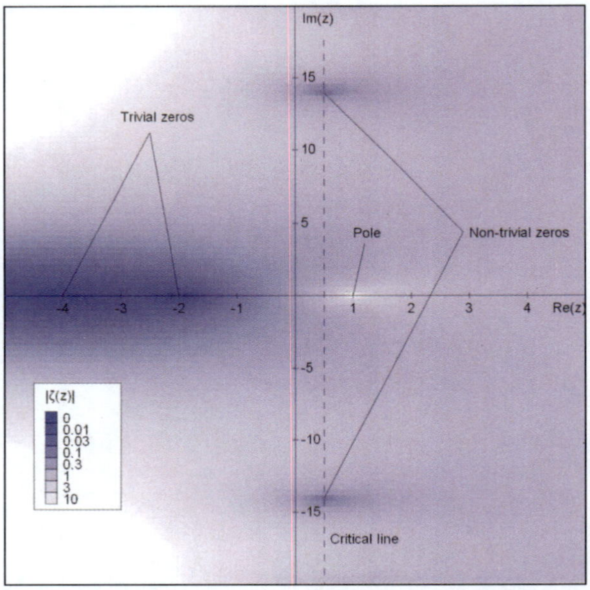

Plot of the absolute values of the zeta function, showing some of its features.

One of the most famous unsolved questions in mathematics, dating from 1859, and one of the Millennium Prize Problems, is the Riemann hypothesis, which asks where the zeros of the Riemann zeta function $\zeta(s)$ are located. This function is an analytic function on the complex numbers. For complex numbers s with real part greater than one it equals both an infinite sum over all integers, and an infinite product over the prime numbers,

$$\zeta(s) = \sum_{n=1}^{\infty} \frac{1}{n^s} = \prod_{p \ prime} \frac{1}{1-p^{-s}}.$$

This equality between a sum and a product, discovered by Euler, is called an Euler product. The Euler product can be derived from the fundamental theorem of arithmetic, and shows the close connection between the zeta function and the prime numbers. It leads to another proof that there are infinitely many primes: if there were only finitely many, then the sum-product equality would also be valid at $s=1$, but the sum would diverge (it is the harmonic series $1+\frac{1}{2}+\frac{1}{3}+\ldots$) while the product would be finite, a contradiction.

The Riemann hypothesis states that the zeros of the zeta-function are all either negative even numbers, or complex numbers with real part equal to 1/2. The original proof of the prime number theorem was based on a weak form of this hypothesis, that there are no zeros with real part equal to 1, although other more elementary proofs have been found. The prime-counting function can be expressed by Riemann's explicit formula as a sum in which each term comes from one of the zeros of the zeta function; the main term of this sum is the logarithmic integral, and the remaining terms cause the sum to fluctuate above and below the main term. In this sense, the zeros control how regularly the prime numbers are distributed. If the Riemann hypothesis is true, these fluctuations will be small, and the asymptotic distribution of primes given by the prime number theorem will also hold over much shorter intervals (of length about the square root of x for intervals near a number x).

Computational Methods

The small gear in this piece of farm equipment has 13 teeth, a prime number, and the middle gear has 21, relatively prime to 13.

For a long time, number theory in general, and the study of prime numbers in particular, was seen as the canonical example of pure mathematics, with no applications outside of mathematics[b] other than the use of prime numbered gear teeth to distribute wear evenly. In particular, number theorists such as British mathematician G. H. Hardy prided themselves on doing work that had absolutely no military significance.

This vision of the purity of number theory was shattered in the 1970s, when it was publicly announced that prime numbers could be used as the basis for the creation of public key cryptography algorithms. These applications have led to significant study of algorithms for computing with prime numbers, and in particular of primality testing, methods for determining whether a given number is prime. The most basic primality testing routine, trial division, is too slow to be useful for large numbers. One group of modern primality tests is applicable to arbitrary numbers, while more efficient tests are available for numbers of special types. Most primality tests only tell whether their argument is prime or not. Routines that also provide a prime factor of composite arguments (or all of its prime factors) are called factorization algorithms. Prime numbers are also used in computing for checksums, hash tables, and pseudorandom number generators.

Trial Division

The most basic method of checking the primality of a given integer n is called *trial division*. This method divides n by each integer from 2 up to the square root of n. Any such integer dividing n evenly establishes n as composite; otherwise it is prime. Integers larger than the square root do not need to be checked because, whenever $n = a \cdot b$, one of the two factors a and b is less than or equal to the square root of n. Another optimization is to check only primes as factors in this range.

For instance, to check whether 37 is prime, this method divides it by the primes in the range from 2 to $\sqrt{37}$, which are 2, 3, and 5. Each division produces a nonzero remainder, so 37 are indeed prime.

Although this method is simple to describe, it is impractical for testing the primality of large integers, because the number of tests that it performs grows exponentially as a function of the number of digits of these integers. However, trial division is still used, with a smaller limit than the square root on the divisor size, to quickly discover composite numbers with small factors, before using more complicated methods on the numbers that pass this filter.

Sieves

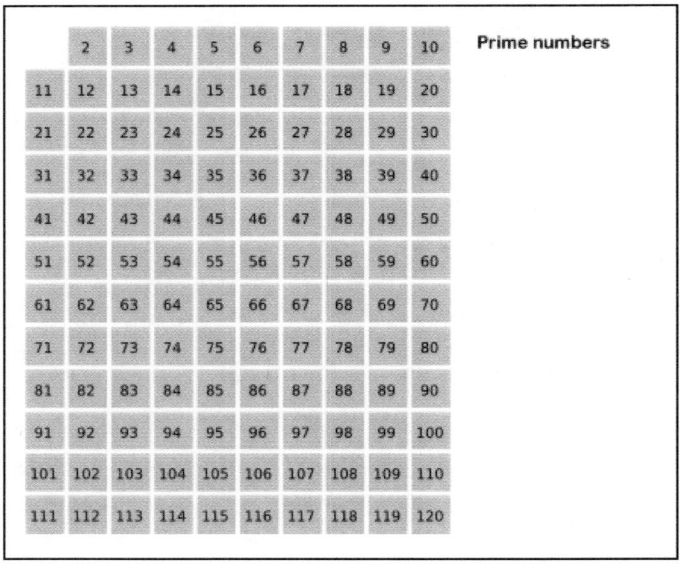

The sieve of Eratosthenes starts with all numbers unmarked (gray). It repeatedly finds the first unmarked number, marks it as prime (dark colors) and marks its square and all later multiples as composite (lighter colors). After marking the multiples of 2 (red), 3 (green), 5 (blue), and 7 (yellow), all primes up to the square root of the table size have been processed, and all remaining unmarked numbers (11, 13, etc.) are marked as primes (magenta).

Before computers, mathematical tables listing all of the primes or prime factorizations up to a given limit were commonly printed. The oldest method for generating a list of primes is called the sieve of Eratosthenes. The animation shows an optimized variant of this method. Another more efficient sieving method for the same problem is the sieve of Atkin. In advanced mathematics, sieve theory applies similar methods to other problems.

Primality Testing versus Primality Proving

Some of the fastest modern tests for whether an arbitrary given number n is prime are probabilistic (or Monte Carlo) algorithms, meaning that they have a small random chance of producing an incorrect answer. For instance the Solovay–Strassen primality test on a given number p chooses a number randomly from 2 through $p-2$ and uses modular exponentiation to check whether $a^{(p-1)/2} \pm 1$ is divisible by p. If so, it answers yes and otherwise it answers no. If p really is prime, it will always answer yes, but if p is composite then it answers yes with probability at most 1/2 and no with probability at least 1/2. If this test is repeated n times on the same number, the probability that a composite number could pass the test every time is at most $1/2^n$. Because this decreases exponentially with the number of tests, it provides high confidence (although not certainty) that a number that passes the repeated test is prime. On the other hand, if the test ever fails, then the number is certainly composite. A composite number that passes such a test is called a pseudoprime.

In contrast, some other algorithms guarantee that their answer will always be correct: primes will always be determined to be prime and composites will always be determined to be composite. For instance, this is true of trial division. The algorithms with guaranteed-correct output include both deterministic (non-random) algorithms, such as the AKS primality test, and randomized Las Vegas algorithms where the random choices made by the algorithm do not affect its final answer, such as some variations of elliptic curve primality proving. When the elliptic curve method concludes that a number is prime, it provides primality certificate that can be verified quickly. The elliptic curve primality test is the fastest in practice of the guaranteed-correct primality tests, but its runtime analysis is based on heuristic arguments rather than rigorous proofs. The AKS primality test has mathematically proven time complexity, but is slower than elliptic curve primality proving in practice. These methods can be used to generate large random prime numbers, by generating and testing random numbers until finding one that is prime; when doing this, a faster probabilistic test can quickly eliminate most composite numbers before a guaranteed-correct algorithm is used to verify that the remaining numbers are prime.

The following table lists some of these tests. Their running time is given in terms of n, the number to be tested and, for probabilistic algorithms, the number k of tests performed. Moreover, k is an arbitrarily small positive number, and log is the logarithm to an unspecified base. The big O notation means that each time bound should be multiplied by a constant factor to convert it from dimensionless units to units of time; this factor depends on implementation details such as the type of computer used to run the algorithm, but not on the input parameters n and k.

Test	Running time
AKS primality test	$O((\log n)^{6+\varepsilon})$
Elliptic curve primality proving	$O((\log n)^{4+\varepsilon})$ heuristically
Baillie-PSW primality test	$O((\log n)^{2+\varepsilon})$
Miller–Rabin primality test	$O(k(\log n)^{2+\varepsilon})$
Solovay–Strassen primality test	$O(k(\log n)^{2+\varepsilon})$

Special-purpose Algorithms and the Largest Known Prime

In addition to the aforementioned tests that apply to any natural number, some numbers of a special form can be tested for primality more quickly. For example, the Lucas–Lehmer primality test can determine whether a Mersenne number (one less than a power of two) is prime, deterministically, in the same time as a single iteration of the Miller–Rabin test. This is why since 1992 the largest *known* prime has always been a Mersenne prime. It is conjectured that there are infinitely many Mersenne primes.

The following table gives the largest known primes of various types. Some of these primes have been found using distributed computing.

Type	Prime	Number of decimal digits
Mersenne prime	$2^{82,589,933} - 1$	24,862,048
Proth prime (but not a Mersenne prime)	$10,223 \times 2^{31,172,165} + 1$	9,383,761
Factorial prime	$208,003! - 1$	1,015,843
Primorial prime[e]	$1,098,133\# - 1$	476,311
Twin primes	$2,996,863,034,895 \times 2^{1,290,000} \pm 1$	388,342

Integer Factorization

Given a composite integer n, the task of providing one (or all) prime factors is referred to as *factorization* of n. It is significantly more difficult than primality testing, and although many factorization algorithms are known, they are slower than the fastest primality testing methods. Trial division and Pollard's rho algorithm can be used to find very small factors of n, and elliptic curve factorization can be effective when n has factors of moderate size. Methods suitable for arbitrary large numbers that do not depend on the size of its factors include the quadratic sieve and general number field sieve. As with primality testing, there are also factorization algorithms that require their input to have a special form, including the special number field sieve. As of January 2018 the largest number known to have been factored by a general-purpose algorithm is RSA-768, which has 232 decimal digits (768 bits) and is the product of two large primes.

Shor's algorithm can factor any integer in a polynomial number of steps on a quantum computer. However, current technology can only run this algorithm for very small numbers. As of October 2012 the largest number that has been factored by a quantum computer running Shor's algorithm is 21.

Other Computational Applications

Several public-key cryptography algorithms, such as RSA and the Diffie–Hellman key exchange, are based on large prime numbers (2048-bit primes are common). RSA relies on the assumption that it is much easier (that is, more efficient) to perform the multiplication of two (large) numbers x and y than to calculate x and y (assumed coprime) if only the product xy is known. The Diffie–Hellman key exchange relies on the fact that there are efficient algorithms for modular exponentiation (computing $a^b \bmod c$), while the reverse operation (the discrete logarithm) is thought to be a hard problem.

Prime numbers are frequently used for hash tables. For instance the original method of Carter and Wegman for universal hashing was based on computing hash functions by choosing random linear functions modulo large prime numbers. Carter and Wegman generalized this method to k-independent hashing by using higher-degree polynomials, again modulo large primes. As well as in the hash function, prime numbers are used for the hash table size in quadratic probing based hash tables to ensure that the probe sequence covers the whole table.

Some checksum methods are based on the mathematics of prime numbers. For instance the checksums used in International Standard Book Numbers are defined by taking the rest of the number modulo 11, a prime number. Because 11 is prime this method can detect both single-digit errors and transpositions of adjacent digits. Another checksum method, Adler-32, uses arithmetic modulo 65521, the largest prime number less than 2^{16}. Prime numbers are also used in pseudorandom number generators including linear congruential generators and the Mersenne Twister.

Other Applications

Prime numbers are of central importance to number theory but also have many applications to other areas within mathematics, including abstract algebra and elementary geometry. For example, it is possible to place prime numbers of points in a two-dimensional grid so that no three are in a line, or so that every triangle formed by three of the points has large area. Another example is Eisenstein's criterion, a test for whether a polynomial is irreducible based on divisibility of its coefficients by a prime number and its square.

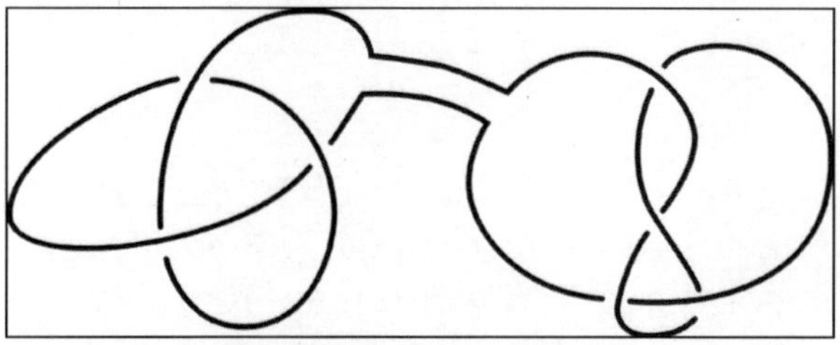

The connected sum of two prime knots.

The concept of prime number is so important that it has been generalized in different ways in various branches of mathematics. Generally, "prime" indicates minimality or indecomposability, in an appropriate sense. For example, the prime field of a given field is its smallest subfield that contains both 0 and 1. It is either the field of rational numbers or a finite field with a prime number of elements, whence the name. Often a second, additional meaning is intended by using the word prime, namely that any object can be, essentially uniquely, decomposed into its prime components. For example, in knot theory, a prime knot is a knot that is indecomposable in the sense that it cannot be written as the connected sum of two nontrivial knots. Any knot can be uniquely expressed as a connected sum of prime knots. The prime decomposition of 3-manifolds is another example of this type.

Beyond mathematics and computing, prime numbers have potential connections to quantum mechanics, and have been used metaphorically in the arts and literature. They have also been used in evolutionary biology to explain the life cycles of cicadas.

Constructible Polygons and Polygon Partitions

Fermat primes are primes of the form

$$F_k = 2^{2^k} + 1,$$

with k a natural number. They are named after Pierre de Fermat, who conjectured that all such numbers are prime. The first five of these numbers – 3, 5, 17, 257, and 65,537 – are prime, but F_5

is composite and so are all other Fermat numbers that have been verified as of 2017. A regular n -gon is constructible using straightedge and compass if and only if the odd prime factors of n (if any) are distinct Fermat primes. Likewise, a regular n -gon may be constructed using straightedge, compass, and an angle trisector if and only if the prime factors of n are any number of copies of 2 or 3 together with a (possibly empty) set of distinct Pierpont primes, primes of the form $2^a 3^b + 1$.

It is possible to partition any convex polygon into n smaller convex polygons of equal area and equal perimeter, when n is a power of a prime number, but this is not known for other values of n.

Quantum Mechanics

Beginning with the work of Hugh Montgomery and Freeman Dyson in the 1970s, mathematicians and physicists have speculated that the zeros of the Riemann zeta function are connected to the energy levels of quantum systems. Prime numbers are also significant in quantum information science, thanks to mathematical structures such as mutually unbiased bases and symmetric informationally complete positive-operator-valued measures.

Biology

The evolutionary strategy used by cicadas of the genus *Magicicada* makes use of prime numbers. These insects spend most of their lives as grubs underground. They only pupate and then emerge from their burrows after 7, 13 or 17 years, at which point they fly about, breed, and then die after a few weeks at most. Biologists theorize that these prime-numbered breeding cycle lengths have evolved in order to prevent predators from synchronizing with these cycles. In contrast, the multi-year periods between flowering in bamboo plants are hypothesized to be smooth numbers, having only small prime numbers in their factorizations.

Prime Factorization

The factorization of a number into its constituent primes, also called prime decomposition. Given a positive integer $n \geq 2$, the prime factorization is written:

$$n = p_1^{a_1} p_2^{a_2} \dots p_k^{a_k},$$

where the p_i s are the k prime factors, each of order a_i. Each factor $p_i^{a_i}$ is called a primary.

Through his invention of the Pratt certificate, Pratt became the first to establish that prime factorization lies in the complexity class NP.

The first few prime factorizations (the number 1, by definition, has a prime factorization of "1") are given in the following table.

n	prime factorization	n	prime factorization
1	1	11	11
2	2	12	$2^2 \cdot 3$
3	3	13	13

4	2^2	14	$2 \cdot 7$
5	5	15	$3 \cdot 5$
6	$2 \cdot 3$	16	2^4
7	7	17	17
8	2^3	18	$2 \cdot 3^2$
9	3^2	19	19
10	$2 \cdot 5$	20	$2^2 \cdot 5$

The number of digits in the prime factorization of $n = 1, 2, \ldots,$ are 1, 1, 1, 2, 1, 2, 1, 2, 2, 2, 2, 3,.

In general, prime factorization is a difficult problem, and many sophisticated prime factorization algorithms have been devised for special types of numbers.

Integers can also be factored over the Gaussian primes. For example, the following table gives the Gaussian integer factorizations for the first few positive integers.

n	factorization
1	1
2	$-i(i+1)^2$
3	3
4	$-(i+1)^4$
5	$-i(2i+1)(2+i)$
6	$-3i(i+1)^2$
7	7
8	$i(i+1)^6$
9	3^2
10	$-(i+1)^2(2i+1)(i+2)$

Interestingly, prime numbers p equal to 1 (mod 4) can always by factored into Gaussian primes in the form:

$${}_1 p = -i(R + I\,i)(I + R\,i),$$

where the real and imaginary parts are inverted in the two parts, while prime numbers equal to 3 (mod 4) cannot be factored into Gaussian primes. This is directly related to Fermat's 4n+1 theorem.

PRIME NUMBER THEOREM

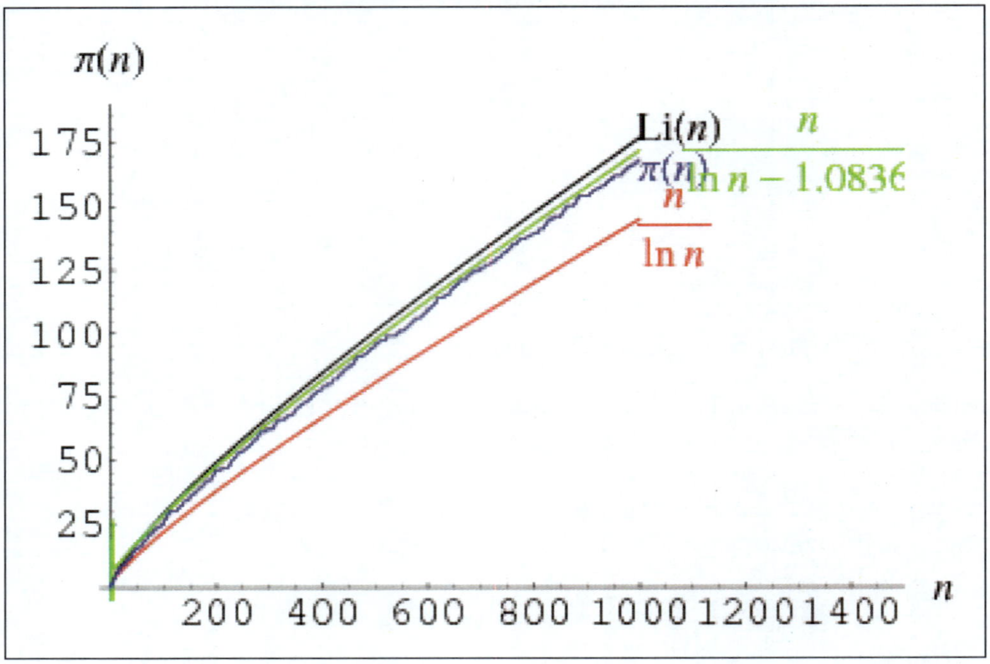

The prime number theorem gives an asymptotic form for the prime counting function $\pi(n)$, which counts the number of primes less than some integer n. For large n,

$$\pi(n) \sim \frac{n}{\mathrm{In}\, n + B}.$$

With $B = -1.08366$ (where B is sometimes called Legendre's constant), a formula which is correct in the leading term only,

$$\frac{n}{\mathrm{In}\, n + B} \sin \frac{n}{\mathrm{In}\, n} - B \frac{n}{(\mathrm{In}\, n)^2} + B^2 \frac{n}{(\mathrm{In}\, n)^3} + \ldots$$

In 1792, when only 15 years old, Gauss proposed that:

$$\pi(n) \sim \frac{n}{\mathrm{In}\, n}.$$

Gauss later refined his estimate to:

$$\pi(n) \sim \mathrm{Li}(n),$$

Where:

$$\mathrm{Li}(n) = \int_2^{\pi} \frac{dx}{\mathrm{In}\, x}$$

is the logarithmic integral. Gauss did not publish this result, which he first mentioned in an 1849 letter to Encke. It was subsequently posthumously published in 1863.

$\mathrm{Li}(n)$ has the asymptotic series about ∞ of:

$$\mathrm{Li}(n) \sim \sum_{k=0}^{\infty} \frac{k!n}{(\mathrm{In}\,n)^{k+1}}$$

$$\sim \frac{n}{\mathrm{In}\,n} + \frac{n}{(\mathrm{In}\,n)^2} + \frac{2n}{(\mathrm{In}\,n)^3} + \ldots,$$

and taking the first three terms has been shown to be a better estimate than $n/\mathrm{In}\,n$ alone The statement $\pi(n) \sim \mathrm{Li}(n)$ is often known as "the" prime number theorem and was proved independently by Hadamard and de la Vallée Poussin. A plot of $\pi(n)$ (lower curve) and $\mathrm{Li}(n)$ is shown above for $n \le 1000$.

For small n, it had been checked and always found that $\pi(n) < \mathrm{Li}(n)$. As a result, many prominent mathematicians, including no less than both Gauss and Riemann, conjectured that the inequality was strict. To everyone's surprise, this conjecture was refuted when Littlewood proved that the inequality reverses infinitely often for sufficiently large n. Skewes then showed that the first crossing of $\pi(n) - \mathrm{Li}(n) = 0$ occurs before $10^{10^{10^{34}}}$, a number now known as the Skewes number. The upper bound for the crossing has subsequently been reduced to 10^{371}. Lehman proved that at least 10^{500} reversals occur for numbers with 1166 or 1167 decimal digits.

Chebyshev put limits on the ratio:

$$\frac{7}{8} < \frac{\pi(n)}{\frac{n}{\mathrm{In}\,n}} < \frac{9}{8}.$$

For large n, he showed that:

$$0.89\,\mathrm{Li}(n) < \pi(n) < 1.11\,\mathrm{Li}(n),$$

Where $\mathrm{Li}(x)$ is the logarithmic integral and

$$0.922M < \frac{\pi(n)}{\frac{n}{\mathrm{In}\,n}} < 1.105.$$

He also showed that if the limit:

$$\lim_{n\to\infty} \frac{\pi(n)}{\frac{n}{\mathrm{In}\,n}}$$

exists, then it is 1. Derbyshire's statement that in 1850, Chebyshev also showed that $\pi(n)$ cannot differ from $n / \ln n$ by more than approximately 10% is therefore correct only for sufficiently large n.

Hadamard and de la Vallée Poussin independently proved the prime number theorem in 1896 by showing that the Riemann zeta function $\zeta(z)$ has no zeros of the form $1+it$, in the sense that no deeper properties of $\zeta(s)$ are required for the proof. Wiener allowed this somewhat vague statement to be interpreted literally, and this proof was simplified by Landau and Bochner.

An elementary proof was found by Erdős and Selberg, although an unfortunate priority dispute over the joint work marred the otherwise beautiful proof. Versions of elementary proofs of the prime number theorem appear in final section of Nagell and in Hardy and Wright. As noted by Hardy and Wright, although this proof is 'elementary,' "this proof is not easy."

Hadamard's proof depends on the simple trigonometric inequality

$$3+4\cos\theta+\cos(2\theta)=2(1+\cos\theta)^2\geq 0.$$

de la Vallée Poussin showed that:

$$\pi(x)=\mathrm{Li}(x)+O\left(\frac{x}{\ln x}e^{-a\sqrt{\ln x}}\right),$$

for some constant a, where $O(x)$ is asymptotic notation. In 1901, Koch showed that if the Riemann hypothesis is true, then

$$\pi(x)=\mathrm{Li}(x)+O\left(\sqrt{x}\ln x\right).$$

which can be written in the slightly weaker form

$$\pi(x)=\mathrm{Li}(x)+O\left(x^{1/2+e}\right).$$

The error term in the above equation has subsequently been improved to:

$$\pi(x)=\mathrm{Li}(x)+O\left(x\exp\left(-\frac{A(\ln x)^{3/5}}{(\ln\ln x)^{1/5}}\right)\right)$$

Ingham proved the prime number theorem using the identity of Ramanujan:

$$\sum_{n=1}^{\infty}\frac{\sigma_a(n)\sigma_b(n)}{n^s}=\frac{\zeta(s)\zeta(s-a)\zeta(s-b)\zeta(s-a-b)}{\zeta(2s-a-b)},$$

where $\sigma_a(n)$ is the divisor function.

Riemann estimated the prime counting function with:

$$\pi(n)\sim\mathrm{Li}(n)-\frac{1}{2}\mathrm{Li}\left(n^{1/2}\right),$$

which is a better approximation than $\mathrm{Li}(n)$ for $n < 10^7$. Riemann also suggested the Riemann function:

$$R(x) = \sum_{n=1}^{\infty} \frac{\mu(n)}{n} \mathrm{Li}\left(x^{1/n}\right),$$

where μ is the Möbius function. An even better approximation for small n (by a factor of 10 for $n < 10^9$) is the Gram series.

The prime number theorem is equivalent to either

$$\lim_{x \to \infty} \frac{\theta(x)}{x} = 1,$$

Or

$$\lim_{x \to \infty} \frac{\psi(x)}{x} = 1,$$

\

Where $\theta(x)$ and $\psi(x)$ are the Chebyshev functions. Chebyshev showed that the only possible limit of these expressions was 1, but was not able to prove existence of the limit.

The Riemann hypothesis is equivalent to the assertion that,

$$\left|\mathrm{Li}(x) - \pi(x)\right| \le c\sqrt{x}\,\mathrm{In}\,x.$$

For some value of c, as shown by Koch in 1901. Some limits obtained without assuming the Riemann hypothesis are.

$$\pi(x) = \mathrm{Li(x)} + O\left[x e^{-(\mathrm{In}\,x)^{1/2}/15}\right]$$

$$\pi(x) = \mathrm{Li(x)} + O\left[x e^{-0.009(\mathrm{In}\,x)^{3/5}/(\mathrm{In}\,\mathrm{In}\,x)^{1/5}}\right].$$

TOTIENT FUNCTION

The totient function $\phi(n)$, also called Euler's totient function, is defined as the number of positive integers $\le n$ that are relatively prime to (i.e., do not contain any factor in common with) n, where 1 is counted as being relatively prime to all numbers. Since a number less than or equal to and relatively prime to a given number is called a totative, the totient function $\phi(n)$ can be simply defined as the number of totatives of n. For example, there are eight totatives of 24 (1, 5, 7, 11, 13, 17, 19, and 23), so $\phi(24) = 8$.

The number $n-\phi(n)$ is called the cototient of n and gives the number of positive integers $\leq n$ that have at least one prime factor in common with n.

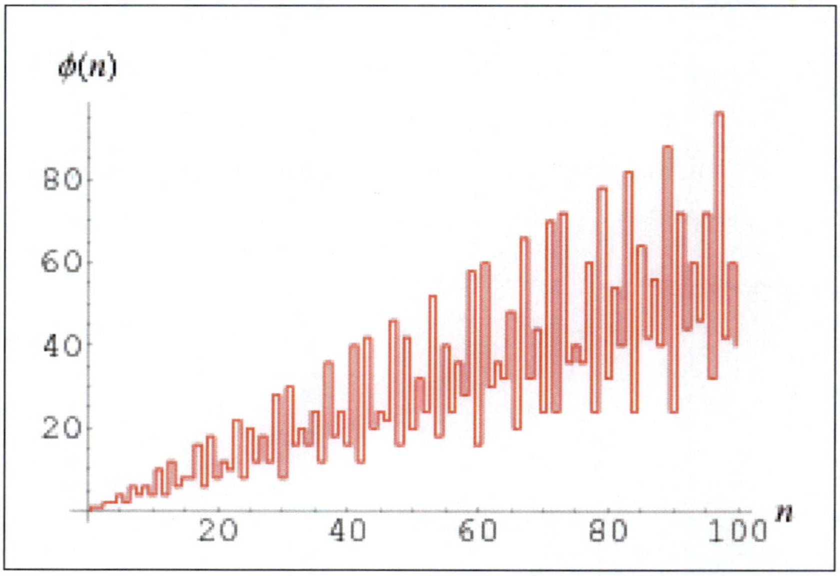

$\phi(n)$ is always even for $n \geq 3$. By convention, $\phi(0)=1$, The first few values of $\phi(n)$ for $n=1,$ 2,... are 1, 1, 2, 2, 4, 2, 6, 4, 6, 4, 10, ... The totient function is given by the Möbius transform of 1, 2, 3, 4,... $\phi(n)$ is plotted above for small n.

For a prime p,

$$\phi(p) = p-1,$$

since all numbers less than p are relatively prime to p. If $m = p^{\alpha}$. is a power of a prime, then the numbers that have a common factor with m are the multiples of $p: p, 2p, \ldots, \left(p^{\alpha-1}\right)p$. There are p^{a-1} of these multiples, so the number of factors relatively prime to p^{α} is

$$\begin{aligned} \phi(p^{\alpha}) &= p^{\alpha} - p^{\alpha-1} \\ &= p^{\alpha-1}(p-1) \\ &= p^{\alpha}\left(1-\frac{1}{p}\right). \end{aligned}$$

Now take a general m divisible by p Let $\phi_p(m)$ be the number of positive integers $\leq m$ not divisible by p As before, $p, 2p, \ldots, (m/p)p$ have common factors, so

$$\begin{aligned} \phi_p(m) &= m - \frac{m}{p} \\ &= m\left(1-\frac{1}{p}\right). \end{aligned}$$

Now let q be some other prime dividing m. The integers divisible by q are $q, 2q, ..., (m/q)q$. But these duplicate $pq,\ 2pq, ..., (m/pq))pq$. So the number of terms that must be subtracted from ϕ_p to obtain ϕ_{pq} is:

$$\begin{aligned}\Delta\phi_q(m) &= \frac{m}{q} - \frac{m}{pq} \\ &= \frac{m}{q}\left(1 - \frac{1}{p}\right),\end{aligned}$$

and,

$$\begin{aligned}\phi_{pq}(m) &= \phi_p(m) - \Delta\phi_q(m) \\ &= m\left(1 - \frac{1}{p}\right) - \frac{m}{q}\left(1 - \frac{1}{p}\right) \\ &= m\left(1 - \frac{1}{p}\right)\left(1 - \frac{1}{q}\right).\end{aligned}$$

By induction, the general case is then:

$$\begin{aligned}\phi(n) &= n\prod_{p|n}\left(1 - \frac{1}{p}\right) \\ &= \left(1 - \frac{1}{p_1}\right)\left(1 - \frac{1}{p_2}\right)...\left(1 - \frac{1}{p_r}\right),\end{aligned}$$

where the product runs over all primes p dividing n. An interesting identity relating $\phi(n^k)$ to $\phi(n)$. is given by

$$\phi(n^k) = n^{k-1}\phi(n)$$

Another identity relates the divisors d of n to n via

$$\sum_{d|n}\phi(d) = n.$$

The totient function is connected to the Möbius function $\mu(n)$ through the sum

$$\sum_{d} d\,\mu\left(\frac{n}{d}\right) = \phi(n),$$

where the sum is over the divisors of n, which can be proven by induction on n and the fact that $\mu(n)$ and $\phi(n)$ are multiplicative.

The totient function has the Dirichlet generating function $\sum_{n=1}^{\infty}\frac{\phi(n)}{n^s}=\frac{\zeta(s-1)}{\zeta(s)}$ for $s>2$.

The totient function satisfies the inequality $\phi(n)\geq\sqrt{n}$ for all n except $n=2$ and $n=6$. Therefore, the only values of n for which $\phi(n)=2$ are $n=3,4$, and 6. In addition, for composite n,

$\phi(n)\leq n-\sqrt{n}$

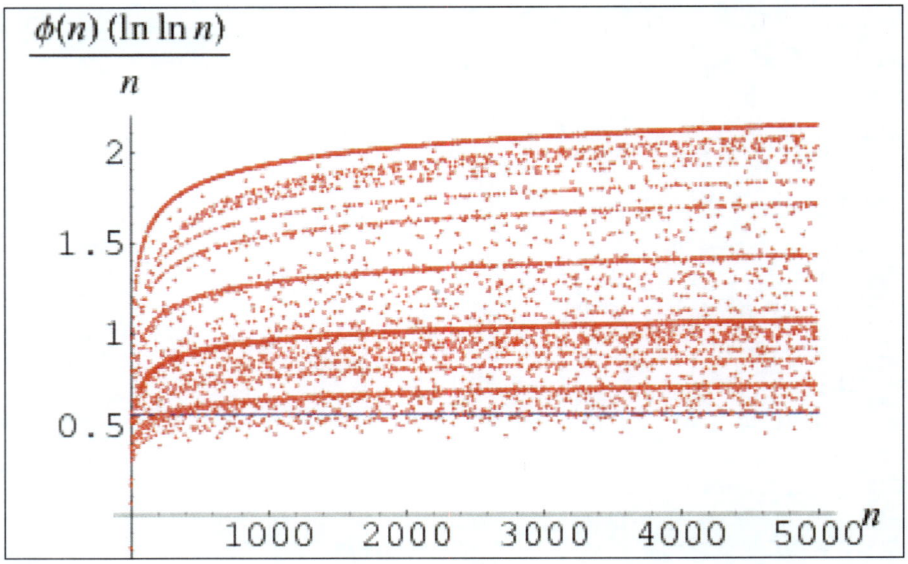

$\phi(n)$ also satisfies:

$$\liminf_{n\to\infty}\phi(n)\frac{\text{In}\,\text{In}\,n}{n}=e^{-\gamma},$$

where γ is the Euler-Mascheroni constant. The values of n for which $\phi(n)<e^{-\gamma}n/(\text{In}\,\text{In}\,n)$ are given by 3, 4, 5, 6, 8, 9, 10, 12, 14, 15, 16, 18, 20, 22,...

The divisor function satisfies the congruence:

$$n\sigma(n)\equiv 2(\text{mod}\,\phi(n))$$
$$=\begin{cases}0(\text{mod}\,\phi(n))\text{ if }\phi(n)=2\\ 2(\text{mod}\,\phi(n))\text{ otherwise}\end{cases}$$

for all primes $p\geq 5$ and no composite with the exception of 4, 6, and 22, where $\sigma(n)$ is the divisor function. This fact was proved by Subbarao, despite the implication to the contrary, “is it true for infinitely many composite n?,” stated in a query subsequently removed from. No composite solution is currently known to $n-1=0(\text{mod}\,\phi(n))$.

A corollary of the Zsigmondy theorem leads to the following congruence, $\phi(a^n+b^n)=0(\text{mod}\,n)$.

The first few n for which $\phi(n)=\phi(n+1)$ are given by 1, 3, 15, 104, 164, 194, 255, 495, 584, 975, ..., which have common values $\phi(n)=1$, 2, 8, 48, 80, 96, 128, 240, 288, 480, ...

The only $n < 10^{10}$ for which

$$\phi(n) = \phi(n+1) = \phi(n+2)$$

is $n = 5186$, giving

$$\phi(5186) = \phi(5187) = \phi(5188) = 2^5\,3^4.$$

Values of $\phi(n)$ shared among n that are close together include:

$$\begin{aligned}\phi(25930) &= \phi(25935) = \phi(25940) = \phi(25942)\\ &= 2^7\,3^4,\\ \phi(404471) &= \phi(404473) = \phi(404477)\\ &= 2^8\,3^2\,5^2\,7.\end{aligned}$$

McCranie found an arithmetic progression of six numbers with equal totient functions,

$$\begin{aligned}&\phi(583200) = \phi(583230) = \phi(583260) = \phi(583290)\\ &= \phi(583320) = \phi(583350) = 155520,\end{aligned}$$

as well as other progressions of six numbers starting at 1166400, 1749600, ...

If the Goldbach conjecture is true, then for every positive integer m, there are primes p and q such that,

$$\phi(p) + \phi(q) = 2m.$$

Erdős asked if this holds for p and q not necessarily prime, but this relaxed form remains unproven discussed solutions to $\phi(\sigma(n)) = n$, where $\sigma(n)$ is the divisor function. F. Helenius has found 365 such solutions, the first of which are 2, 8, 12, 128, 240, 720, 6912, 32768, 142560, 712800, ...

TWIN PRIME

A twin prime is a prime number that is either 2 less or 2 more than another prime number—for example, either member of the twin prime pair (41, 43). In other words, a twin prime is a prime that has a prime gap of two. Sometimes the term *twin prime* is used for a pair of twin primes; an alternative name for this is prime twin or prime pair.

Twin primes become increasingly rare as one examines larger ranges, in keeping with the general tendency of gaps between adjacent primes to become larger as the numbers themselves get larger. However, it is unknown whether there are infinitely many twin primes or there is a largest pair. The work of Yitang Zhang in 2013, as well as work by James Maynard, Terence Tao and others, has made substantial progress towards proving that there are infinitely many twin primes, but at present this remains unsolved.

Properties

Usually the pair (2, 3) is not considered to be a pair of twin primes. Since 2 is the only even prime, this pair is the only pair of prime numbers that differ by one; thus twin primes are as closely spaced as possible for any other two primes.

The first few twin prime pairs are: (3, 5), (5, 7), (11, 13), (17, 19), (29, 31), (41, 43), (59, 61), (71, 73), (101, 103), (107, 109), (137, 139), ….

Five is the only prime in two distinct pairs.

Every twin prime pair except (3,5) is of the form $(6n-1, 6n+1)$ for some natural number n; that is, the number between the two primes is a multiple of 6. As a result, the sum of any pair of twin primes (other than 3 and 5) is divisible by 12.

In 1915, Viggo Brun showed that the sum of reciprocals of the twin primes was convergent. This famous result, called Brun's theorem, was the first use of the Brun sieve and helped initiate the development of modern sieve theory. The modern version of Brun's argument can be used to show that the number of twin primes less than N does not exceed

$$\frac{CN}{(\log N)^2}$$

for some absolute constant $C > 0$. In fact, it is bounded above by: $\frac{C'N}{(\log N)^2}\left(1+O\left(\frac{\log\log N}{\log N}\right)\right)$, where $C' = 8C_2$, where C_2 is the twin prime constant, given below.

Other Theorems Weaker than the Twin Prime Conjecture

In 1940, Paul Erdős showed that there is a constant $c < 1$ and infinitely many primes p such that $(p' - p) < (c \ln p)$ where p' denotes the next prime after p. What this means is that we can find infinitely many intervals that contain two primes (p,p') as long as we let these intervals grow slowly in size as we move to bigger and bigger primes. Here, "grow slowly" means that the length of these intervals can grow logarithmically. This result was successively improved; in 1986 Helmut Maier showed that a constant $c < 0.25$ can be used. In 2004 Daniel Goldston and Cem Yıldırım showed that the constant could be improved further to $c = 0.085786\ldots$ In 2005, Goldston, János Pintz and Yıldırım established that c can be chosen to be arbitrarily small, i.e.

$$\liminf_{n\to\infty} \frac{p_{n+1} - p_n}{\log p_n} = 0.$$

On the other hand, this result does not rule out that there may not be infinitely many intervals that contain two primes if we only allow the intervals to grow in size as, for example, $c \ln \ln p$.

By assuming the Elliott–Halberstam conjecture or a slightly weaker version, they were able to show that there are infinitely many n such that at least two of n, $n + 2$, $n + 6$, $n + 8$, $n + 12$, $n + 18$, or $n + 20$ are prime. Under a stronger hypothesis they showed that for infinitely many n, at least two of n, $n + 2$, $n + 4$, and $n + 6$ are prime.

The result of Yitang Zhang,

$$\liminf_{n\to\infty}(p_{n+1}-p_n)<N \; with \; N=7\times10^7,$$

is a major improvement on the Goldston–Graham–Pintz–Yıldırım result. The Polymath Project optimization of Zhang's bound and the work of Maynard has reduced the bound to $N = 246$.

Conjectures

First Hardy–Littlewood Conjecture

The Hardy–Littlewood conjecture (named after G. H. Hardy and John Littlewood) is a generalization of the twin prime conjecture. It is concerned with the distribution of prime constellations, including twin primes, in analogy to the prime number theorem. Let $\pi_2(x)$ denote the number of prime's $p \le x$ such that $p + 2$ is also prime. Define the twin prime constant C_2 as:

$$C_2=\prod_{\substack{p\ prime \\ p\ge3}}\left(1-\frac{1}{(p-1)^2}\right)\approx 0.66016181584686957392781211001 4....$$

(here the product extends over all prime numbers $p \ge 3$). Then a special case of the first Hardy-Littlewood conjecture is that:

$$\pi_2(x)\sim 2C_2\frac{x}{(\ln x)^2}\sim 2C_2\int_2^x\frac{dt}{(\ln t)^2}.$$

in the sense that the quotient of the two expressions tends to 1 as x approaches infinity. (The second ~ is not part of the conjecture and is proven by integration by parts.)

The conjecture can be justified (but not proven) by assuming that $1 / \ln t$ describes the density function of the prime distribution, an assumption suggested by the prime number theorem and would imply the twin prime conjecture, but remains unresolved.

The fully general first Hardy–Littlewood conjecture on prime k-tuples (not given here) implies that the *second* Hardy–Littlewood conjecture is false.

Polignac's Conjecture

Polignac's conjecture from 1849 states that for every positive even natural number k, there are infinitely many consecutive prime pairs p and p' such that $p' - p = k$ (i.e. there are infinitely many prime gaps of size k). The case $k = 2$ is the twin prime conjecture. The conjecture has not yet been proven or disproven for any specific value of k, but Zhang's result proves that it is true for at least one (currently unknown) value of k. Indeed, if such a k did not exist, then for any positive even natural number N there are at most finitely many n such that $p_{n+1} - p_n = m$ for all $m < N$ and so for n large enough we have $p_{n+1} - p_n > N$, which would contradict Zhang's result.

Large Twin Primes

Beginning in 2007, two distributed computing projects, Twin Prime Search and PrimeGrid, have produced several record-largest twin primes. As of September 2018, the current largest twin prime pair known is $2996863034895 \cdot 2^{1290000} \pm 1$, with 388,342 decimal digits. It was discovered in September 2016.

There are 808,675,888,577,436 twin prime pairs below 10^{18}.

An empirical analysis of all prime pairs up to $4.35 \cdot 10^{15}$ shows that if the number of such pairs less than x is $f(x) \cdot x/(\log x)^2$ then $f(x)$ is about 1.7 for small x and decreases towards about 1.3 as x tends to infinity. The limiting value of $f(x)$ is conjectured to equal twice the twin prime constant, according to the Hardy–Littlewood conjecture.

Other Elementary Properties

Every third odd number is divisible by 3, which requires that no three successive odd numbers can be prime unless one of them is 3. Five is therefore the only prime that is part of two twin prime pairs. The lower member of a pair is by definition a Chen prime.

It has been proven that the pair $(m, m + 2)$ is a twin prime if and only if

$$4((m-1)!+1) \equiv -m \left(\bmod\, m(m+2)\right).$$

If $m - 4$ or $m + 6$ is also prime then the three primes are called a prime triplet.

For a twin prime pair of the form $(6n - 1, 6n + 1)$ for some natural number $n > 1$, n must have units digit 0, 2, 3, 5, 7, or 8.

Isolated Prime

An isolated prime (also known as single prime or non-twin prime) is a prime number p such that neither $p - 2$ nor $p + 2$ is prime. In other words, p is not part of a twin prime pair. For example, 23 is an isolated prime, since 21 and 25 are both composite.

The first few isolated primes are 2, 23, 37, 47, 53, 67, 79, 83, 89, 97, ...

It follows from Brun's theorem that almost all primes are isolated in the sense that the ratio of the number of isolated primes less than a given threshold n and the number of all primes less than n tends to 1 as n tends to infinity.

Brun's Theorem

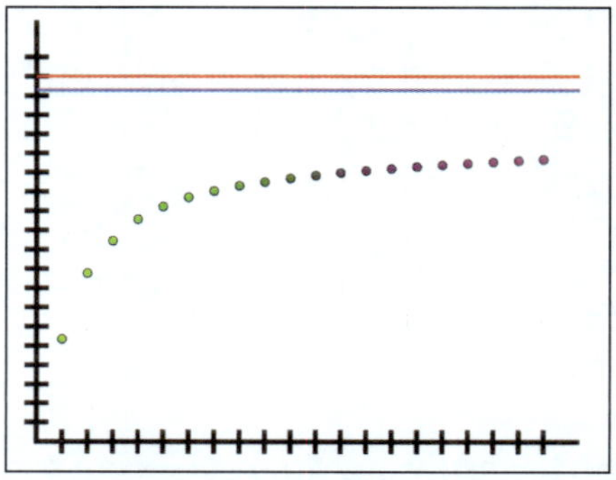

The convergence to Brun's constant.

In number theory, Brun's theorem states that the sum of the reciprocals of the twin primes (pairs of prime numbers which differ by 2) converges to a finite value known as Brun's constant, usually denoted by B_2. Brun's theorem was proved by Viggo Brun in 1919, and it has historical importance in the introduction of sieve methods.

Asymptotic Bounds on Twin Primes

The convergence of the sum of reciprocals of twin primes follows from bounds on the density of the sequence of twin primes. Let $\pi_2(x)$ denote the number of primes $p \le x$ for which $p + 2$ is also prime (i.e. $\pi_2(x)$ is the number of twin primes with the smaller at most x). Then, for $x \ge 3$, we have:

$$\pi_2(x) = O\left(\frac{x(\log\log x)^2}{(\log x)^2}\right).$$

That is, twin primes are less frequent than prime numbers by nearly a logarithmic factor. It follows from this bound that the sum of the reciprocals of the twin primes converges, or stated in other words, the twin primes form a small set. In explicit terms the sum:

$$\sum_{p:p+2\in\mathbb{P}}\left(\frac{1}{p}+\frac{1}{p+2}\right)=\left(\frac{1}{3}+\frac{1}{5}\right)+\left(\frac{1}{5}+\frac{1}{7}\right)+\left(\frac{1}{11}+\frac{1}{13}\right)+\cdots.$$

either has finitely many terms or has infinitely many terms but is convergent: Its value is known as Brun's constant.

The fact that the sum of the reciprocals of the prime numbers diverges implies that there are infinitely many prime numbers. Because the sum of the reciprocals of the twin primes instead converges, it is not possible to conclude from this result that there are finitely many or infinitely many twin primes. Brun's constant could be an irrational number only if there are infinitely many twin primes.

Numerical Estimates

The series converges extremely slowly. Thomas Nicely remarks that after summing the first one billion (10^9) terms, the relative error is still more than 5%..

By calculating the twin primes up to 10^{14} (and discovering the Pentium FDIV bug along the way), Nicely heuristically estimated Brun's constant to be 1.902160578. Nicely has extended his computation to 1.6×10^{15} as of 18 January 2010 but this is not the largest computation of its type.

In 2002, Pascal Sebah and Patrick Demichel used all twin primes up to 10^{16} to give the estimate that $B_2 \approx 1.902160583104$. Hence,

Year	B_2	# of twin primes used
1976	1.902160540	1×10^{11}
1996	1.902160578	1×10^{14}
2002	1.902160583104	1×10^{16}

The last is based on extrapolation from the sum 1.830484424658... for the twin primes below 1016. Dominic Klyve showed conditionally (in an unpublished thesis) that B2 < 2.1754 (assuming the extended Riemann hypothesis). It has been shown unconditionally that B2 < 2.347.

There is also a Brun's constant for prime quadruplets. A prime quadruplet is a pair of two twin prime pairs, separated by a distance of 4 (the smallest possible distance). The first prime quadruplets are (5, 7, 11, 13), (11, 13, 17, 19), (101, 103, 107, 109). Brun's constant for prime quadruplets, denoted by B_4, is the sum of the reciprocals of all prime quadruplets:

$$B_4 = \left(\frac{1}{5}+\frac{1}{7}+\frac{1}{11}+\frac{1}{13}\right)+\left(\frac{1}{11}+\frac{1}{13}+\frac{1}{17}+\frac{1}{19}\right)+\left(\frac{1}{101}+\frac{1}{103}+\frac{1}{107}+\frac{1}{109}\right)+\cdots$$

with value:

B_4 = 0.87058 83800 ± 0.00000 00005, the error range having a 99% confidence level according to Nicely.

This constant should not be confused with the Brun's constant for cousin primes, as prime pairs of the form $(p, p + 4)$, which is also written as B_4. Wolf derived an estimate for the Brun-type sums B_n of $4/n$.

Further Results

Let $C_2 = 0.6601\ldots$ be the twin prime constant. Then it is conjectured that,

$$\pi_2(x) \sim 2C_2 \frac{x}{(\log x)^2}.$$

In particular,

$$\pi_2(x) < (2C_2 + \varepsilon)\frac{x}{(\log x)^2}$$

for every $\varepsilon > 0$ and all sufficiently large x.

Many special cases of the above have been proved. Most recently, Jie Wu proved that for sufficiently large x,

$$\pi_2(x) < 4.5\frac{x}{(\log x)^2}$$

where 4.5 corresponds to $\varepsilon \approx 3.18$ in the above.

COUSIN PRIME

Pairs of primes of the form $(p, p+4)$ are called cousin primes. The first few are (3, 7), (7, 11), (13, 17), (19, 23), (37, 41), (43, 47), (67, 71),...

A large pair of cousin (proven) primes start with:

$$p = 9771919142 \cdot [(53238 \cdot 7879\#)^2 - 1] + 2310 \cdot 53238 \cdot 7879\# / 385 + 1,$$

where $7879\#$ is a primorial. These primes have 10154 digits and were found by T. Alm, M. Fleuren, and J. K. Andersen.

As of Jan. 2006, the largest known pair of cousin (probable) primes are

$$630062 \cdot 2^{37555} + 3, 7,$$

which have 11311 digits and were found by D. Johnson in May 2004.

According to the first Hardy-Littlewood conjecture, the cousin primes have the same asymptotic density as the twin primes,

$$P\,(p, p \quad 4) \sim 2\prod_{\geq} \frac{p(p \quad)}{(p \quad)^2} \int \frac{dx'}{(\mathrm{In}\, x')^2}$$

$$\sim 1.320323632\ldots \int \frac{dx}{(\mathrm{In} \quad ')}$$

where $\prod_2 = 1.320323632\ldots$ is the twin primes constant.

An analogy to Brun's constant, the constant

$$B_4 \equiv \left(\frac{1}{7} + \frac{1}{11}\right) + \left(\frac{1}{13} + \frac{1}{17}\right) + \left(\frac{1}{19} + \frac{1}{23}\right) + \left(\frac{1}{37} + \frac{1}{41}\right) + \ldots,$$

(omitting the initial term $1/3 + 1/7$) can be defined. Using cousin primes up to 2^{42}, the value of B_4 is estimated as $B_4 \approx 1.1970449$.

PRIME TRIPLET

In mathematics, a prime triplet is a set of three prime numbers in which the smallest and largest of the three differ by 6. In particular, the sets must have the form (*p*, *p* + 2, *p* + 6) or (*p*, *p* + 4, *p* + 6). With the exceptions of (2, 3, 5) and (3, 5, 7), this is the closest possible grouping of three prime numbers, since one of every three sequential odd numbers is a multiple of three, and hence not prime (except for 3 itself).

Examples: The first prime triplets are

(5, 7, 11), (7, 11, 13), (11, 13, 17), (13, 17, 19), (17, 19, 23), (37, 41, 43), (41, 43, 47), (67, 71, 73), (97, 101, 103), (101, 103, 107), (103, 107, 109), (107, 109, 113), (191, 193, 197), (193, 197, 199), (223, 227, 229), (227, 229, 233), (277, 281, 283), (307, 311, 313), (311, 313, 317), (347, 349, 353), (457, 461,

463), (461, 463, 467), (613, 617, 619), (641, 643, 647), (821, 823, 827), (823, 827, 829), (853, 857, 859), (857, 859, 863), (877, 881, 883), (881, 883, 887).

Subpairs of Primes

A prime triplet contains a pair of twin primes (p and $p + 2$, or $p + 4$ and $p + 6$), a pair of cousin primes (p and $p + 4$, or $p + 2$ and $p + 6$), and a pair of sexy primes (p and $p + 6$).

Higher-order Versions

A prime can be a member of up to three prime triplets - for example, 103 is a member of (97, 101, 103), (101, 103, 107) and (103, 107, 109). When this happens, the five involved primes form a prime quintuplet.

A prime quadruplet (p, $p + 2$, $p + 6$, $p + 8$) contains two overlapping prime triplets, (p, $p + 2$, $p + 6$) and ($p + 2$, $p + 6$, $p + 8$).

Conjecture on Prime Triplets

Similarly to the twin prime conjecture, it is conjectured that there are infinitely many prime triplets. The first known gigantic prime triplet was found in 2008 by Norman Luhn and François Morain. The primes are (p, $p + 2$, $p + 6$) with $p = 2072644824759 \times 2^{33333} - 1$. As of August 2018 the largest known prime triplet contains primes with 16737 digits and was found by Peter Kaiser. The primes are (p, $p + 4$, $p + 6$) with $p = 6521953289619 \times 2^{55555} - 5$.

The Skewes number for the triplet (p, $p + 2$, $p + 6$) is 87613571, and for the triplet (p, $p + 4$, $p + 6$) it is 337867.

PRIME QUADRUPLET

Prime Quadruplet is a prime constellation of four successive primes with minimal distance $(p, p+2, p+6, p+8)$. The quadruplet (2, 3, 5, 7) has smaller minimal distance, but it is an exceptional special case. With the exception of (5, 7, 11, 13), a prime quadruple must be of the form $(30n+11, 30n+13, 30n+17, 30n+19)$. The first few values of n which give prime quadruples are $n = 0$, 3, 6, 27, 49, 62, 69, 108, 115,... and the first few values of p are 5 (the exceptional case), 11, 101, 191, 821, 1481, 1871, 2081, 3251, 3461,... The number of prime quadruplets with largest member less than $10^1, 10^2, \ldots$, are 1, 2, 5, 12, 38, 166, 899, 4768,... .

The asymptotic formula for the frequency of prime quadruples is analogous to that for other prime constellations,

$$P_x(p, p+2, p+6, p+8) \sim \frac{27}{2} \prod_{p \geq 5} \frac{p^3(p-4)}{(p-1)^4} \int_2^x \frac{dx}{(\ln x)^4}$$

$$= 4.151180864 \int_2^x \frac{dx}{(\ln x)^4},$$

where $c = 4.15118...$ is the Hardy-Littlewood constant for prime quadruplets.

Roonguthai found the large prime quadruplets with

$$p = 10^{99} + 349781731$$
$$p = 10^{199} + 21156403891$$
$$p = 10^{299} + 140159459341$$
$$p = 10^{399} + 34993836001$$
$$p = 10^{499} + 883750143961$$
$$p = 10^{599} + 1394283756151$$
$$p = 10^{699} + 547634621251.$$

Forbes found the large quadruplet with

$$p = 76912895956636885(2^{3279} - 2^{1093}) - 6 \cdot 2^{1093} - 7.$$

GOLDBACH'S CONJECTURE

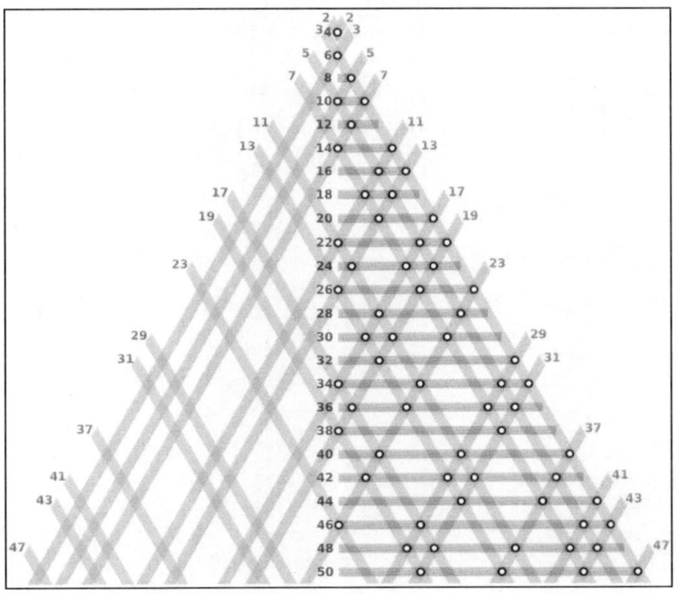

The even integers from 4 to 28 as sums of two primes: Even integers correspond to horizontal lines. For each prime, there are two oblique lines, one red and one blue. The sums of two primes are the intersections of one red and one blue line, marked by a circle. Thus the circles on a given horizontal line give all partitions of the corresponding even integer into the sum of two primes.

Goldbach's conjecture is one of the oldest and best-known unsolved problems in number theory and all of mathematics. It states:

> Every even integer greater than 2 can be expressed as the sum of two primes.

The conjecture has been shown to hold for all integers less than 4×10^{18}, but remains unproven despite considerable effort.

Goldbach Number

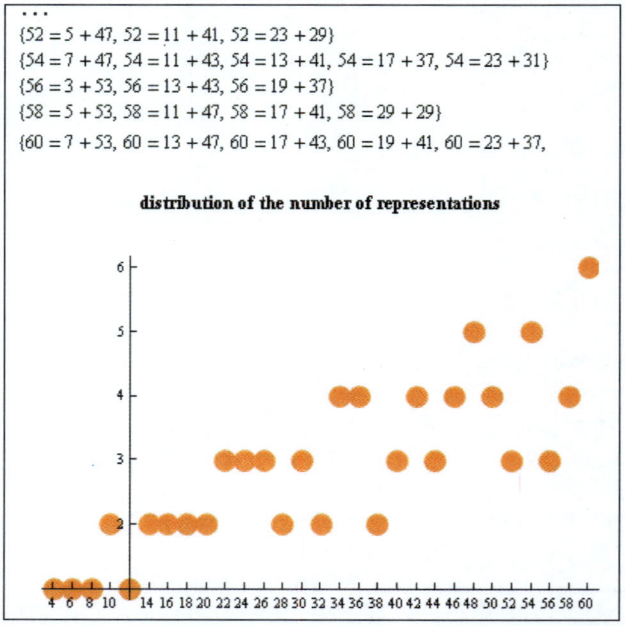

The number of ways an even number can be represented as the sum of two primes.

A Goldbach number is a positive even integer that can be expressed as the sum of two odd primes. Since 4 is the only even number greater than 2 that requires the even prime 2 in order to be written as the sum of two primes, another form of the statement of Goldbach's conjecture is that all even integers greater than 4 are Goldbach numbers.

The expression of a given even number as a sum of two primes is called a Goldbach partition of that number. The following are examples of Goldbach partitions for some even numbers:

$6 = 3 + 3$

$8 = 3 + 5$

$10 = 3 + 7 = 5 + 5$

$12 = 7 + 5$

...

$100 = 3 + 97 = 11 + 89 = 17 + 83 = 29 + 71 = 41 + 59 = 47 + 53$

...

The number of ways in which $2n$ can be written as the sum of two primes (for n starting at 1) is:

0, 1, 1, 1, 2, 1, 2, 2, 2, 2, 3, 3, 3, 2, 3, 2, 4, 4, 2, 3, 4, 3, 4, 5, 4, 3, 5, 3, 4, 6, 3, 5, 6, 2, 5, 6, 5, 5, 7, 4, 5, 8, 5, 4, 9, 4, 5, 7, 3, 6, 8, 5, 6, 8, 6, 7, 10, 6, 6, 12, 4, 5, 10, 3, ...

Verified Results

For small values of n, the strong Goldbach conjecture (and hence the weak Goldbach conjecture) can be verified directly. For instance, Nils Pipping in 1938 laboriously verified the conjecture up to $n \le 10^5$. With the advent of computers, many more values of n have been checked; T. Oliveira e Silva ran a distributed computer search that has verified the conjecture for $n \le 4 \times 10^{18}$ (and double-checked up to 4×10^{17}) as of 2013. One record from this search is that 3325581707333960528 is the smallest number that has no Goldbach partition with a prime below 9781.

Heuristic Justification

Statistical considerations that focus on the probabilistic distribution of prime numbers present informal evidence in favour of the conjecture (in both the weak and strong forms) for sufficiently large integers: the greater the integer, the more ways there are available for that number to be represented as the sum of two or three other numbers, and the more "likely" it becomes that at least one of these representations consists entirely of primes.

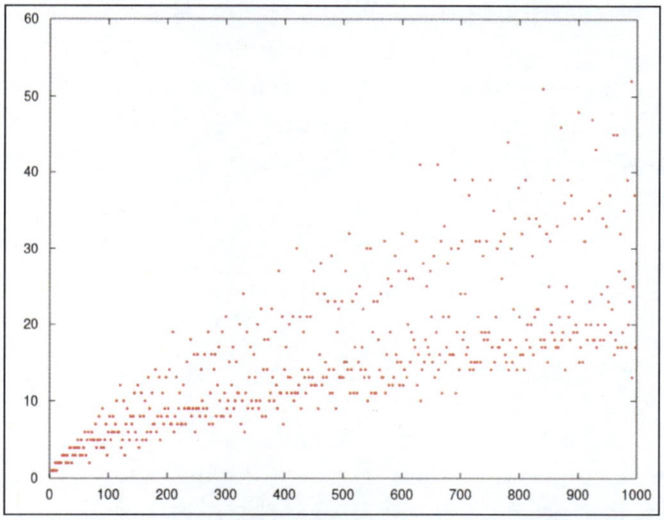

Number of ways to write an even number n as the sum of two primes (4 ≤ n ≤ 1,000).

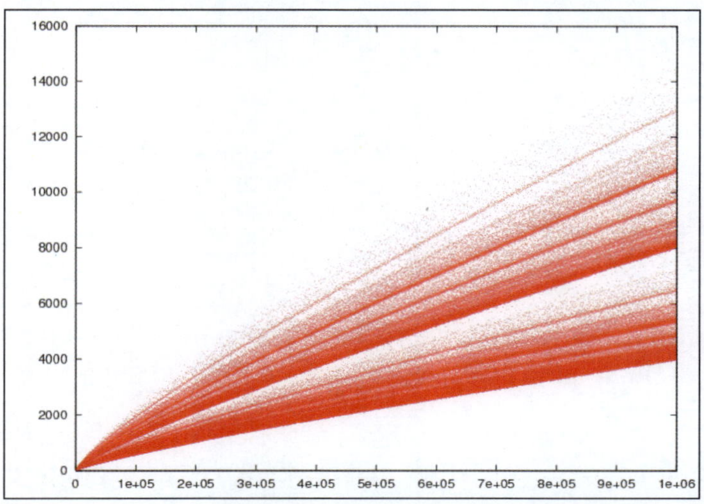

Number of ways to write an even number n as the sum of two primes (4 ≤ n ≤ 1000000).

A very crude version of the heuristic probabilistic argument (for the strong form of the Goldbach conjecture) is as follows. The prime number theorem asserts that an integer m selected at random has roughly a $1/\ln m$ chance of being prime. Thus if n is a large even integer and m is a number between 3 and $n/2$, then one might expect the probability of m and $n - m$ simultaneously being prime to be $1/\left[\ln m \ln(n-m)\right]$. If one pursues this heuristic, one might expect the total number of ways to write a large even integer n as the sum of two odd primes to be roughly

$$\sum_{m=3}^{n/2} \frac{1}{\ln m} \frac{1}{\ln(n-m)} \approx \frac{n}{2(\ln n)^2}.$$

Since this quantity goes to infinity as n increases, we expect that every large even integer has not just one representation as the sum of two primes, but in fact has very many such representations.

This heuristic argument is actually somewhat inaccurate, because it assumes that the events of m and $n - m$ being prime are statistically independent of each other. For instance, if m is odd, then $n - m$ is also odd, and if m is even, then $n - m$ is even, a non-trivial relation because, besides the number 2, only odd numbers can be prime. Similarly, if n is divisible by 3, and m was already a prime distinct from 3, then $n - m$ would also be coprime to 3 and thus be slightly more likely to be prime than a general number. Pursuing this type of analysis more carefully, Hardy and Littlewood in 1923 conjectured that for any fixed $c \geq 2$, the number of representations of a large integer n as the sum of c primes $n = p_1 + \cdots + p_c$ with $p_1 \leq \cdots \leq p_c$ should be asymptotically equal to,

$$\left(\prod_p \frac{p\gamma_{c,p}(n)}{(p-1)^c}\right) \int_{2 \leq x_1 \leq \cdots \leq x_c : x_1 + \cdots + x_c = n} \frac{dx_1 \cdots dx_{c-1}}{\ln x_1 \cdots \ln x_c},$$

where the product is over all primes p, and $\gamma_{c,p}(n)$ is the number of solutions to the equation $n = q_1 + \cdots + q_c \mod p$ in modular arithmetic, subject to the constraints $q_1, \ldots, q_c \neq 0 \mod p$. This formula has been rigorously proven to be asymptotically valid for $c \geq 3$ from the work of Vinogradov, but is still only a conjecture when $c = 2$. In the latter case, the above formula simplifies to 0 when n is odd, and to

$$2\Pi_2 \left(\prod_{p|n;\, p\geq 3} \frac{p-1}{p-2}\right) \int_2^n \frac{dx}{(\ln x)^2} \approx 2\Pi_2 \left(\prod_{p|n;\, p\geq 3} \frac{p-1}{p-2}\right) \frac{n}{(\ln n)^2},$$

when n is even, where Π_2 is Hardy–Littlewood's twin prime constant

$$\Pi_2 := \prod_{p\geq 3} \left(1 - \frac{1}{(p-1)^2}\right) = 0.6601618158\ldots.$$

This is sometimes known as the extended Goldbach conjecture. The strong Goldbach conjecture is in fact very similar to the twin prime conjecture, and the two conjectures are believed to be of roughly comparable difficulty.

The Goldbach partition functions shown here can be displayed as histograms, which informatively illustrate the above equations.

Rigorous Results

The strong Goldbach conjecture is much more difficult than the weak Goldbach conjecture. Using Vinogradov's method, Chudakov, Van der Corput, and Estermann showed that almost all even numbers can be written as the sum of two primes (in the sense that the fraction of even numbers which can be so written tends towards 1). In 1930, Lev Schnirelmann proved that any natural number greater than 1 can be written as the sum of not more than C prime numbers, where C is an effectively computable constant, see Schnirelmann density. Schnirelmann's constant is the lowest number C with this property. Schnirelmann himself obtained $C < 800\,000$. This result was subsequently enhanced by many authors, such as Olivier Ramaré, who in 1995 showed that every even number $n \geq 4$ is in fact the sum of at most 6 primes. The best known result currently stems from the proof of the weak Goldbach conjecture by Harald Helfgott, which directly implies that every even number $n \geq 4$ is the sum of at most 4 primes.

In 1924 Hardy and Littlewood showed under the assumption of the generalized Riemann hypothesis that the amount of even numbers up to X violating the Goldbach conjecture is much less than $X^{(1/2)+c}$ for small c.

Chen Jingrun showed in 1973 using the methods of sieve theory that every sufficiently large even number can be written as the sum of either two primes, or a prime and a semiprime (the product of two primes).

In 1975, Hugh Montgomery and Robert Charles Vaughan showed that "most" even numbers are expressible as the sum of two primes. More precisely, they showed that there exist positive constants c and C such that for all sufficiently large numbers N, every even number less than N is the sum of two primes, with at most CN^{1-c} exceptions. In particular, the set of even integers that are not the sum of two primes has density zero.

In 1951, Linnik proved the existence of a constant K such that every sufficiently large even number is the sum of two primes and at most K powers of 2. Roger Heath-Brown and Jan-Christoph Schlage-Puchta in 2002 found that $K = 13$ works.

As with many famous conjectures in mathematics, there are a number of purported proofs of the Goldbach conjecture, none of which are accepted by the mathematical community.

Related Problems

Although Goldbach's conjecture implies that every positive integer greater than one can be written as a sum of at most three primes, it is not always possible to find such a sum using a greedy algorithm that uses the largest possible prime at each step. The Pillai sequence tracks the numbers requiring the largest number of primes in their greedy representations.

One can consider similar problems in which primes are replaced by other particular sets of numbers, such as the squares.

- It was proven by Lagrange that every positive integer is the sum of four squares.
- Hardy and Littlewood listed as their Conjecture I: "Every large odd number ($n > 5$) is the sum of a prime and the double of a prime" (Mathematics Magazine, 66.1: 45–47). This conjecture is known as Lemoine's conjecture (also called Levy's conjecture).

- The Goldbach conjecture for practical numbers, a prime-like sequence of integers, was stated by Margenstern in 1984, and proved by Melfi in 1996: every even number is a sum of two practical numbers.

BATEMAN–HORN CONJECTURE

In number theory, the Bateman–Horn conjecture is a statement concerning the frequency of prime numbers among the values of a system of polynomials, named after mathematicians Paul T. Bateman and Roger A. Horn, of The University of Utah, who proposed it in 1962. It provides a vast generalization of such conjectures as the Hardy and Littlewood conjecture on the density of twin primes or their conjecture on primes of the form $n^2 + 1$; it is also a strengthening of Schinzel's hypothesis H.

The Bateman–Horn conjecture provides a conjectured density for the positive integers at which a given set of polynomials all have prime values. For a set of m distinct irreducible polynomials $f_1, \ldots, f_m$ with integer coefficients, an obvious necessary condition for the polynomials to simultaneously generate prime values infinitely often is that they satisfy Bunyakovsky's property, that there does not exist a prime number p that divides their product $f(n)$ for every positive integer n. For, if there were such a prime p, having all values of the polynomials simultaneously prime for a given n would imply that at least one of them must be equal to p, which can only happen for finitely many values of n or there would be a polynomial with infinitely many roots, whereas the conjecture is how to give conditions where the values are simultaneously prime for infinitely many n.

An integer n is prime-generating for the given system of polynomials if every polynomial $f_i(n)$ produces a prime number when given n as its argument. If $P(x)$ is the number of prime-generating integers among the positive integers less than x, then the Bateman–Horn conjecture states that,

$$P(x) \sim \frac{C}{D} \int_2^x \frac{dt}{(\log t)^m},$$

where D is the product of the degrees of the polynomials and where C is the product over primes p

$$C = \prod_p \frac{1 - N(p)/p}{(1 - 1/p)^m},$$

with $N(p)$ the number of solutions to $f(n) \equiv 0 \pmod{p}$.

Bunyakovsky's property implies $N(p) < p$ for all primes p, so each factor in the infinite product C is positive. Intuitively one then naturally expects that the constant C is itself positive, and with some work this can be proved. (Work is needed since some infinite products of positive numbers equal zero).

Negative Numbers

As stated above, the conjecture is not true: the single polynomial $f_1(x) = -x$ produces only negative numbers when given a positive argument, so the fraction of prime numbers among its values is always zero. There are two equally valid ways of refining the conjecture to avoid this difficulty:

- One may require all the polynomials to have positive leading coefficients, so that only a constant number of their values can be negative.
- Alternatively, one may allow negative leading coefficients but count a negative number as being prime when its absolute value is prime.

It is reasonable to allow negative numbers to count as primes as a step towards formulating more general conjectures that apply to other systems of numbers than the integers, but at the same time it is easy to just negate the polynomials if necessary to reduce to the case where the leading coefficients are positive.

Examples:

If the system of polynomials consists of the single polynomial $f_1(x) = x$, then the values n for which $f_1(n)$ is prime are themselves the prime numbers, and the conjecture becomes a restatement of the prime number theorem.

If the system of polynomials consists of the two polynomials $f_1(x) = x$ and $f_2(x) = x + 2$, then the values of n for which both $f_1(n)$ and $f_2(n)$ are prime are just the smaller of the two primes in every pair of twin primes. In this case, the Bateman–Horn conjecture reduces to the Hardy–Littlewood conjecture on the density of twin primes, according to which the number of twin prime pairs less than x is,

$$\pi_2(x) \sim 2\prod_{p\geq 3}\frac{p(p-2)}{(p-1)^2}\frac{x}{(\log x)^2} \approx 1.32\frac{x}{(\log x)^2}.$$

Analogue for Polynomials Over a Finite Field

When the integers are replaced by the polynomial ring $F[u]$ for a finite field F, one can ask how often a finite set of polynomials $f_i(x)$ in $F[u][x]$ simultaneously takes irreducible values in $F[u]$ when we substitute for x elements of $F[u]$. Well-known analogies between integers and $F[u]$ suggest an analogue of the Bateman–Horn conjecture over $F[u]$, but the analogue is wrong. For example, data suggest that the polynomial $x^3 + u$ in $F_3[u][x]$ takes(asymptotically) the expected number of irreducible values when x runs over polynomials in $F_3[u]$ of odd degree, but it appears to take (asymptotically) twice as many irreducible values as expected when x runs over polynomials of degree that is 2 mod 4, while it (provably) takes *no* irreducible values at all when x runs over nonconstant polynomials with degree that is a multiple of 4. An analogue of the Bateman–Horn conjecture over $F[u]$ which fits numerical data uses an additional factor in the asymptotics which depends on the value of d mod 4, where d is the degree of the polynomials in $F[u]$ over which x is sampled.

WARING'S PROBLEM

In number theory, Waring's problem asks whether each natural number k has an associated positive integer s such that every natural number is the sum of at most s natural numbers to the power of k. For example, every natural number is the sum of at most 4 squares, 9 cubes, or 19 fourth powers. Waring's problem was proposed in 1770 by Edward Waring, after whom it is named. Its affirmative answer, known as the Hilbert–Waring theorem, was provided by Hilbert in 1909. Waring's problem has its own Mathematics Subject Classification, 11P05, "Waring's problem and variants."

Relationship with Lagrange's Four-square Theorem

Long before Waring posed his problem, Diophantus had asked whether every positive integer could be represented as the sum of four perfect squares greater than or equal to zero. This question later became known as Bachet's conjecture, after the 1621 translation of Diophantus by Claude Gaspard Bachet de Méziriac, and it was solved by Joseph-Louis Lagrange in his four-square theorem in 1770, the same year Waring made his conjecture. Waring sought to generalize this problem by trying to represent all positive integers as the sum of cubes, integers to the fourth power, and so forth, to show that any positive integer may be represented as the sum of other integers raised to a specific exponent, and that there was always a maximum number of integers raised to a certain exponent required to represent all positive integers in this way.

The Number $g(k)$

For every k, let $g(k)$ denote the minimum number s of kth powers of naturals needed to represent all positive integers. Every positive integer is the sum of one first power, itself, so $g(1)=1$. Some simple computations show that 7 requires 4 squares, 23 requires 9 cubes, and 79 requires 19 fourth-powers; these examples show that $g(2)\geq 4$, $g(3)\geq 9$, and $g(4)\geq 19$. Waring conjectured that these lower bounds were in fact exact values.

Lagrange's four-square theorem of 1770 states that every natural number is the sum of at most four squares. Since three squares are not enough, this theorem establishes $g(2)=4$. Lagrange's four-square theorem was conjectured in Bachet's 1621 edition of Diophantus's Arithmetica; Fermat claimed to have a proof, but did not publish it.

Over the years various bounds were established, using increasingly sophisticated and complex proof techniques. For example, Liouville showed that $g(4)$ is at most 53. Hardy and Littlewood showed that all sufficiently large numbers are the sum of at most 19 fourth powers.

That $g(3)=9$ was established by Wieferich and A. J. Kempner, $g(4)=19$ by R. Balasubramanian, F. Dress, and J.-M. Deshouillers, $g(5)=37$ in 1964 by Chen Jingrun, and $g(6)=73$ in 1940 by Pillai.

Let $\lfloor x \rfloor$ and $\{x\}$ respectively denote the integral and fractional part of a positive real number x. Since $c=2^k\lfloor (3/2)^k \rfloor -1<3^k$, only 2^k and 1^k can be used to represent this number c; the most economical representation requires $\lfloor (3/2)^k \rfloor -1$ terms of 2^k and 2^k-1 terms of 1^k. It follows that $g(k)$ is at least as large as $2^k+\lfloor (3/2)^k \rfloor -2$. This was noted by J. A. Euler, the son of Leonhard Euler, in about 1772. Later work by Dickson, Pillai, Rubugunday, Niven and many others has proved that

$$g(k)=2^k+\lfloor (3/2)^k \rfloor -2 \quad \text{if } 2^k\{(3/2)^k\}+\lfloor (3/2)^k \rfloor \leq 2^k$$

$$g(k)=2^k+\lfloor (3/2)^k \rfloor+\lfloor (4/3)^k \rfloor -2 \quad \text{if } 2^k\{(3/2)^k\}+\lfloor (3/2)^k \rfloor > 2^k \text{ and } \lfloor (4/3)^k \rfloor\lfloor (3/2)^k \rfloor+\lfloor (4/3)^k \rfloor+\lfloor (3/2)^k \rfloor = 2^k$$

$$g(k)=2^k+\lfloor (3/2)^k] \rfloor+\lfloor (4/3)^k \rfloor -3 \quad \text{if } 2^k\{(3/2)^k\}+\lfloor (3/2)^k \rfloor > 2^k \text{ and } \lfloor (4/3)^k \rfloor\lfloor (3/2)^k \rfloor+\lfloor (4/3)^k \rfloor+\lfloor (3/2)^k \rfloor > 2^k.$$

No value of k is known for which $2^k\{(3/2)^k\}+\lfloor(3/2)^k\rfloor>2^k$. Mahler proved there can only be a finite number of such k, and Kubina and Wunderlich have shown that any such k must satisfy $k>471{,}600{,}000$. Thus it is conjectured that this never happens, that is, $g(k)=2^k+\lfloor(3/2)^k\rfloor-2$ for every positive integer k.

The first few values of $g(k)$ are:

1, 4, 9, 19, 37, 73, 143, 279, 548, 1079, 2132, 4223, 8384, 16673, 33203, 66190, 132055, 263619, 526502, 1051899, ...

The Number *G*(*k*)

From the work of Hardy and Littlewood, the related quantity $G(k)$ was studied with $g(k)$. $G(k)$ is defined to be the least positive integer s such that every sufficiently large integer (i.e. every integer greater than some constant) can be represented as a sum of at most s positive integers to the power of k. Clearly, $G(1) = 1$. Since squares are congruent to 0, 1, or 4 (mod 8), no integer congruent to 7 (mod 8) can be represented as a sum of three squares, implying that $G(2) \geq 4$. Since $G(k) \leq g(k)$ for all k, this shows that $G(2) = 4$. Davenport showed that $G(4) = 16$ in 1939, by demonstrating that any sufficiently large number congruent to 1 through 14 mod 16 could be written as a sum of 14 fourth powers. The exact value of $G(k)$ is unknown for any other k, but there exist bounds.

Lower Bounds for *G*(*k*)

The number $G(k)$ is greater than or equal to:

2^{r+2} if $k = 2^r$ with $r \geq 2$, or $k = 3 \times 2^r$;

p^{r+1} if p is a prime greater than 2 and $k = p^r(p-1)$;

$(p^{r+1}-1)/2$ if p is a prime greater than 2 and $k = p^r(p-1)/2$;

$k+1$ for all integers k greater than 1.

In the absence of congruence restrictions, a density argument suggests that $G(k)$ should equal $k + 1$.

Bounds
$1 = G(1) = 1$
$4 = G(2) = 4$
$4 \leq G(3) \leq 7$
$16 = G(4) = 16$
$6 \leq G(5) \leq 17$
$9 \leq G(6) \leq 24$
$8 \leq G(7) \leq 33$
$32 \leq G(8) \leq 42$
$13 \leq G(9) \leq 50$
$12 \leq G(10) \leq 59$
$12 \leq G(11) \leq 67$

$16 \le G(12) \le 76$
$14 \le G(13) \le 84$
$15 \le G(14) \le 92$
$16 \le G(15) \le 100$
$64 \le G(16) \le 109$
$18 \le G(17) \le 117$
$27 \le G(18) \le 125$
$20 \le G(19) \le 134$
$25 \le G(20) \le 142$

Upper Bounds for *G*(*k*)

G(3) is at least four (since cubes are congruent to 0, 1 or −1 mod 9); for numbers less than 1.3×10^9, 1290740 is the last to require six cubes, and the number of numbers between N and 2N requiring five cubes drops off with increasing N at sufficient speed to have people believe that G(3) = 4; the largest number now known not to be a sum of four cubes is 7373170279850, and the authors give reasonable arguments there that this may be the largest possible. The upper bound G(3) ≤ 7 is due to Linnik in 1943. (All nonnegative integers require at most 9 cubes, and the largest integers requiring 9, 8, 7, 6 and 5 cubes are conjectured to be 239, 454, 8042, 1290740 and 7373170279850, respectively).

13792 is the largest number to require seventeen fourth powers (Deshouillers, Hennecart and Landreau showed in 2000 that every number between 13793 and 10^{245} required at most sixteen, and Kawada, Wooley and Deshouillers extended Davenport's 1939 result to show that every number above 10^{220} required no more than sixteen). Sixteen fourth powers are always needed to write a number of the form $31\cdot16^n$.

617597724 is the last number less than 1.3×10^9 which requires ten fifth powers, and 51033617 the last number less than 1.3×10^9 which requires eleven.

The upper bounds on the right with $k = 5, 6, \ldots, 20$ are due to Vaughan and Wooley.

Using his improved Hardy-Littlewood method, I. M. Vinogradov published numerous refinements leading to,

$$G(k) \le k(3\log k + 11),$$

in 1947 and, ultimately,

$$G(k) \le k(2\log k + 2\log\log k + C\log\log\log k),$$

for an unspecified constant *C* and sufficiently large *k* in 1959.

Applying his p-adic form of the Hardy-Littlewood-Ramanujan-Vinogradov method to estimating trigonometric sums, in which the summation is taken over numbers with small prime divisors, Anatolii Alexeevitch Karatsuba obtained a new estimate of the Hardy function $G(k)$ (for $(k \ge 400)$:

$$G(k) < 2k\log k + 2k\log\log k + 12k.$$

Further refinements were obtained by Vaughan.

Wooley then established that for some constant C,

$$G(k) \leq k \log k + k \log \log k + Ck.$$

HARDY–LITTLEWOOD CIRCLE METHOD

In mathematics, the Hardy–Littlewood circle method is a technique of analytic number theory. It is named for G. H. Hardy and J. E. Littlewood, who developed it in a series of papers on Waring's problem.

The goal is to prove asymptotic behavior of a series: to show that $a_n \sim F(n)$ for some function. This is done by taking the generating function of the series, then computing the residues about zero (essentially the Fourier coefficients). Technically, the generating function is scaled to have radius of convergence 1, so it has singularities on the unit circle – thus one cannot take the contour integral over the unit circle.

The circle method is specifically how to compute these residues, by partitioning the circle into minor arcs (the bulk of the circle) and major arcs (small arcs containing the most significant singularities), and then bounding the behavior on the minor arcs. The key insight is that, in many cases of interest (such as theta functions), the singularities occur at the roots of unity, and the significance of the singularities is in the order of the Farey sequence. Thus one can investigate the most significant singularities, and, if fortunate, compute the integrals.

Setup

The circle in question was initially the unit circle in the complex plane. Assuming the problem had first been formulated in the terms that for a sequence of complex numbers,

$$a_n, n = 0, 1, 2, 3, \ldots$$

we want some asymptotic information of the type,

$$a_n \sim F(n)$$

where we have some heuristic reason to guess the form taken by F (an ansatz), we write,

$$f(z) = \sum a_n z^n$$

a power series generating function. The interesting cases are where f is then of radius of convergence equal to 1, and we suppose that the problem as posed has been modified to present this situation.

Residues

From that formulation, it follows directly from the residue theorem that,

$$I_n = \int f(z) z^{-(n+1)}\, dz = 2\pi i a_n,$$

for integers $n \geq 0$, where the integral is taken over the circle of radius r and centred at 0, for any r with,

$$0 < r < 1.$$

That is, this is a contour integral, with the contour being the circle described traversed once anti-clockwise. So far, this is relatively elementary. We would like to take $r = 1$ directly, i.e. to use the unit circle contour. In the complex analysis formulation this is problematic, since the values of f are not in general defined there.

Singularities on Unit Circle

The problem addressed by the circle method is to force the issue of taking $r = 1$, by a good understanding of the nature of the singularities f exhibits on the unit circle. The fundamental insight is the role played by the Farey sequence of rational numbers, or equivalently by the roots of unity

$$\zeta = \exp\left(\frac{2\pi i r}{s}\right).$$

Here the denominator s, assuming that r/s is in lowest terms, turns out to determine the relative importance of the singular behaviour of typical f near ζ.

Method

The Hardy–Littlewood circle method, for the complex-analytic formulation, can then be thus expressed. The contributions to the evaluation of I_n, as $r \to 1$, should be treated in two ways, traditionally called major arcs and minor arcs. We divide the roots of unity ζ into two classes, according to whether $s \leq N$, or $s > N$, where N is a function of n that is ours to choose conveniently. The integral I_n is divided up into integrals each on some arc of the circle that is adjacent to ζ, of length a function of s (again, at our discretion). The arcs make up the whole circle; the sum of the integrals over the *major arcs* is to make up $2\pi i F(n)$ (realistically, this will happen up to a manageable remainder term). The sum of the integrals over the *minor arcs* is to be replaced by an upper bound, smaller in order than $F(n)$.

Applications

Refinements of the method have allowed results to be proved about the solutions of homogeneous Diophantine equations, as long as the number of variables k is large relative to the degree d. This turns out to be a contribution to the Hasse principle, capable of yielding quantitative information. If d is fixed and k is small, other methods are required, and indeed the Hasse principle tends to fail.

Rademacher's Contour

In the special case when the circle method is applied to find the coefficients of a modular form of negative weight, Hans Rademacher found a modification of the contour that makes the series arising from the circle method converge to the exact result. To describe his contour, it is convenient to replace the unit circle by the upper half plane, by making the substitution $z = \exp(2\pi i\tau)$, so that the contour integral becomes an integral from $\tau = i$ to $\tau = 1 + i$. (The number i could be replaced by any number on the upper half plane, but i is the most convenient choice.) Rademacher's contour is (more or less) given by the boundaries of all the Ford circles from 0 to 1, as shown in the diagram. The replacement of the line from i to $1 + i$ by the boundaries of these circles is a non-trivial limiting process, which can be justified for modular forms that have negative weight, and with more care can also be justified for non-constant terms for the case of weight 0 (in other words modular functions).

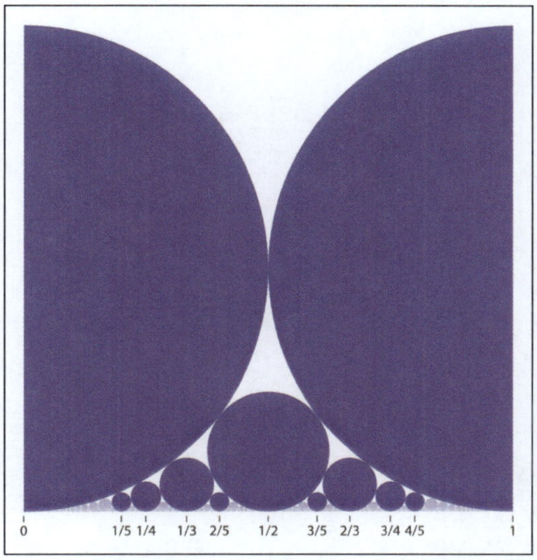

Ford circles: A circle rests upon each fraction in lowest terms. The darker circles shown are for the fractions 0/1, 1/1, 1/2, 1/3, 2/3, 1/4, 3/4, 1/5, 2/5, 3/5 and 4/5. Each circle is tangential to the base line and its neighboring circles. Fractions with the same denominator have circles of the same size.

SECOND HARDY–LITTLEWOOD CONJECTURE

In number theory, the second Hardy–Littlewood conjecture concerns the number of primes in intervals.

The conjecture states that,

$$\pi(x + y) \le \pi(x) + \pi(y),$$

for $x, y \ge 2$, where $\pi(x)$ denotes the prime-counting function, giving the number of prime numbers up to and including x.

This means that the number of primes from $x + 1$ to $x + y$ is always less than or equal to the number of primes from 1 to y. This was proved to be inconsistent with the first Hardy–Littlewood conjecture on prime k-tuples, and the first violation is expected to likely occur for very large values of x. For example, an admissible k-tuple (or prime constellation) of 447 primes can be found in an interval of $y = 3159$ integers, while $\pi(3159) = 446$. If the first Hardy–Littlewood conjecture holds, then the first such k-tuple is expected for x greater than 1.5×10^{174} but less than 2.2×10^{1198}.

SIERPINSKI NUMBER

In number theory, a Sierpiński number is an odd natural number k such that $k \times 2^n + 1$ is composite, for all natural numbers n. In 1960, Wacław Sierpiński proved that there are infinitely many odd integers k which have this property.

In other words, when k is a Sierpiński number, all members of the following set are composite:

$$\left\{k \cdot 2^n + 1 : n \in \mathbb{N}\right\}.$$

If the form instead is $k \times 2^n - 1$, then k is a Riesel number.

Known Sierpiński Numbers

The sequence of currently *known* Sierpiński numbers begins with:

78557, 271129, 271577, 322523, 327739, 482719, 575041, 603713, 903983, 934909, 965431, 1259779, 1290677, 1518781, 1624097, 1639459, 1777613, 2131043, 2131099, 2191531, 2510177, 2541601, 2576089, 2931767, 2931991, ...

The number 78557 was proved to be a Sierpiński number by John Selfridge in 1962, who showed that all numbers of the form $78557 \cdot 2^n + 1$ have a factor in the covering set {3, 5, 7, 13, 19, 37, 73}. For another known Sierpiński number, 271129, the covering set is {3, 5, 7, 13, 17, 241}. Most currently known Sierpiński numbers possess similar covering sets.

However, in 1995 A. S. Izotov showed that some fourth powers could be proved to be Sierpiński numbers without establishing a covering set for all values of n. His proof depends on the aurifeuillean factorization $t^4 \cdot 2^{4m+2} + 1 = (t^2 \cdot 2^{2m+1} + t \cdot 2^{m+1} + 1) \cdot (t^2 \cdot 2^{2m+1} - t \cdot 2^{m+1} + 1)$. This establishes that all $n \equiv 2 \pmod 4$ give rise to a composite, and so it remains to eliminate only $n \equiv 0, 1, 3 \pmod 4$ using a covering set.

Sierpiński Problem

The Sierpiński problem asks for the value of the smallest Sierpiński number. In 1967, Sierpiński and Selfridge conjectured that 78,557 is the smallest Sierpiński number, and thus the answer to the Sierpiński problem.

To show that 78,557 really is the smallest Sierpiński number, one must show that all the odd numbers smaller than 78,557 are *not* Sierpiński numbers. That is, for every odd k below 78,557, there needs to exist a positive integer n such that $k2^n + 1$ is prime. As of November 2018, there are only five candidates which have not been eliminated as possible Sierpiński numbers:

k = 21181, 22699, 24737, 55459, and 67607.

The most recently eliminated candidate was k = 10223, when the prime $10223 \times 2^{31172165} + 1$ was discovered by PrimeGrid. This number is 9,383,761 digits long.

Prime Sierpiński Problem

In 1976, Nathan Mendelsohn determined that the second provable Sierpiński number is the prime $k = 271129$. The prime Sierpiński problem asks for the value of the smallest *prime* Sierpiński number, and there is an ongoing "Prime Sierpiński search" which tries to prove that 271129 is the first Sierpiński number which is also a prime. The prime values of k less than 271129 for which a prime of the form $k2^n + 1$ is not known are

k = 22699, 67607, 79309, 79817, 152267, 156511, 222113, 225931, and 237019.

The first two, being less than 78557, are also unsolved cases of the (non-prime) Sierpiński problem described above. The most recently eliminated candidate was $k = 168451$, when the prime number $168451 \times 2^{19375200} + 1$ was discovered by PrimeGrid. The number is 5,832,522 digits long.

Extended Sierpiński Problem

Suppose that both preceding Sierpiński problems had finally been solved, showing that 78557 is the smallest Sierpiński number and that 271129 is the smallest prime Sierpiński number. This still leaves unsolved the question of the *second* Sierpinski number; there could exist a composite Sierpiński number k such that $78557 < k < 271129$. An ongoing search is trying to prove that 271129 is the second Sierpiński number, by testing all k values between 78557 and 271129, prime or not.

Solving the extended Sierpiński problem, the most demanding of the three posed problems, requires the elimination of 23 remaining candidates $k < 271129$, of which nine are prime and fourteen are composite. The latter include k = 21181, 24737, 55459 from the original Sierpiński problem, unique to the extended Sierpiński problem. The following ten values of k remain:

k = 91549, 99739, 131179, 163187, 200749, 202705, 209611, 227723, 229673, and 238411.

The distributed volunteer computing project PrimeGrid is attempting to eliminate all the remaining values of k. As of September 2019, no prime has been found for these values of k with $n < 13081160$.

The most recent elimination was in April 2018, when $193997 \times 2^{11452891} + 1$ was found to be prime by PrimeGrid, eliminating k = 193997. The number is 3,447,670 digits long.

Simultaneously Sierpiński and Riesel

A number may be simultaneously Sierpiński and Riesel. These are called Brier numbers. The smallest five known examples are 3316923598096294713661, 10439679896374780276373, 11615103277955704975673, 12607110588854501953787, 17855036657007596110949, ...

Dual Sierpinski Problem

If we take n to be a negative integer, then the number k2n + 1 becomes $\frac{2^{|n|} + k}{2^{|n|}}$. When k is odd, this is a fraction in reduced form, with numerator 2|n| + k. A dual Sierpinski number is defined as an odd natural number k such that 2 n + k is composite for all natural numbers n. There is a conjecture that the set of these numbers is the same as the set of Sierpinski numbers; for example, 2n + 78557 is composite for all natural numbers n.

For odd values of k the least n such that $2^n + k$ is prime are,

1, 1, 1, 2, 1, 1, 2, 1, 1, 2, 1, 3, 2, 1, 1, 4, 2, 1, 2, 1, 1, 2, 1, 5, 2, ...

The odd values of k for which $2^n + k$ is composite for all $n < k$ are,

773, 2131, 2491, 4471, 5101, 7013, 8543, 10711, 14717, 17659, 19081, 19249, 20273, 21661, 22193, 26213, 28433, ...

NIVEN'S CONSTANT

Given a positive integer $m > 1$, let its prime factorization be written $m = p_1^{a_1} p_2^{a_2} p_3^{a_3} \ldots p_k^{a_k}$.

Define the functions $h(n)$ and $H(n)$ by $h(1) = 1, H(1) = 1$, and

$$h(m) = \min(a_1, a_2, \ldots, a_k)$$
$$H(m) = \max(a_1, a_2, \ldots, a_k).$$

The first few terms of $h(m)$ are 1, 1, 1, 2, 1, 1, 1, 3, 2, 1, 1, 1, 1, 1, 1, 4, ... while the first few terms of $H(m)$ are 1, 1, 1, 2, 1, 1, 1, 3, 2, 1, 1, 2, 1, 1, 1, 4, ...

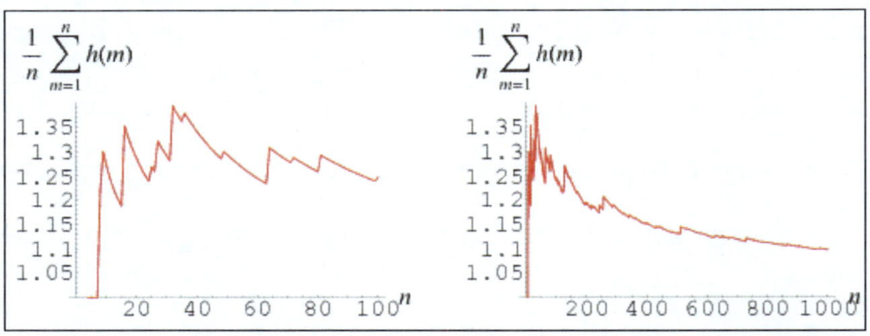

Then the average value of $h(m)$ tends to,

$$\lim_{n\to\infty} \frac{1}{n} \sum_{m=1}^{n} h(m) = 1.$$

Here, the running average values are given by 1/2, 2/3, 3/4, 1, 1, 1, 1, 11/9, 13/10, 14/11, 5/4, 16/13, ...

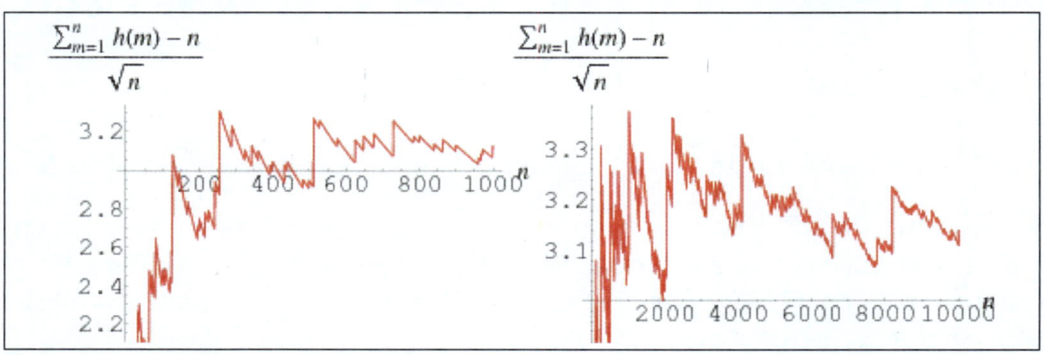

In addition, the ratio,

$$\lim_{n\to\infty}\frac{\sum_{m=1}^{n} h(m)-n}{\sqrt{n}}=\frac{\zeta\left(\frac{3}{2}\right)}{\zeta(3)},$$

Where $\zeta(z)$ is the Riemann zeta function.

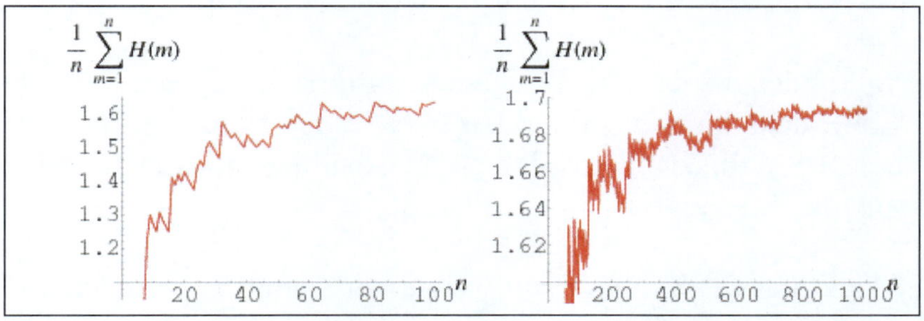

Niven also proved that,

$$\lim_{n\to\infty}\frac{1}{n}\sum_{m=1}^{n} H(m)=C,$$

where Niven's constant C is given by,

$$C=1\left\{\sum_{j=2}^{\infty}\left[1-\frac{1}{\zeta(j)}\right]\right\}=1.705211\ldots$$

Here, the running average values are given by 1/2, 2/3, 3/4, 1, 1, 1, 1, 11/9, 13/10, 14/11, 5/4, 17/13,...

The continued fraction of Niven's constant is 1, 1, 2, 2, 1, 1, 4, 1, 1, 3, 4, 4, 8, 4, 1,... The positions at which the digits 1, 2,... first occur in the continued fraction are 1, 3, 10, 7, 47, 41, 34, 13, 140, 252, 20, ... The sequence of largest terms in the continued fraction is 1, 2, 4, 8, 11, 14, 29, 372, 559,... which occur at positions 1, 3, 7, 13, 20, 35, 51, 68, 96, ...

References

- Stillwell, John (2010). Mathematics and Its History. Undergraduate Texts in Mathematics (3rd ed.). Springer. P. 40. ISBN 978-1-4419-6052-8
- Analytic Number Theory: wolfram.com, Retrieved 19 April, 2019
- Pieprzyk, Josef; Hardjono, Thomas; Seberry, Jennifer (2013). "2.3.9 Probabilistic Computations". Fundamentals of Computer Security. Springer. Pp. 51–52. ISBN 978-3-662-07324-7
- Primefactorization: wolfram.com, Retrieved 5 February, 2019
- Mckee, Maggie (14 May 2013). "First proof that infinitely many prime numbers come in pairs". Nature. Doi:10.1038/nature.2013.12989. ISSN 0028-0836
- Tomás Oliveira e Silva (7 April 2008). "Tables of values of pi(x) and of pi2(x)". Aveiro University. Retrieved 7 January 2011
- Cousinprimes, wolfram.com, Retrieved 26 July, 2019

Algebraic Number Theory

CHAPTER 4

The concepts that fall under the domain of algebraic number of algebraic number theory include Gaussian integer, Cubic field, biquadratic field, roots of unity, class field theory, ideal class group, etc. The topics elaborated in this chapter will help in providing a thorough understanding of these concepts of algebraic number theory.

Algebraic Number Theory is the branch of number theory with the basic aim of studying properties of algebraic integers in algebraic number fields K of finite degree over the field Q of rational numbers.

The set of algebraic integers O_K of a field K/O an extension K of Q of degree n – can be obtained from an integral basis $(\omega_1,...,\omega_n)$; this means that each algebraic integer (i.e. element of O_K) can be written in the form $x_1\,\omega_1+...+x_n\,\omega_n$ where all the x_i run through the rational integers (i.e. z).

Moreover, such a representation is unique for each algebraic integer in K.

However, properties of rational integers often do not have obvious analogues for algebraic integers. The first such property is related to units, the invertible elements of O_K. The field of rational numbers has only +1 and −1 as units, but a general algebraic number field may contain an infinite number of units. E.g. consider the real quadratic field $Q(\sqrt{D})$, where $D>1$ is a rational integer not equal to a square. If, moreover, $D\not\equiv 1(\text{mod}\,4)$, then $(1,\sqrt{D})$ is an integral basis for it. The Pell equation $x^2-D\text{y}^2=1$ has an infinite number of solutions $(x,\text{y})\in Z\times Z$. Any of them gives rise to a unit $x+\text{y}\sqrt{D}$ and $Q(\sqrt{D})$. In fact,

$$(x+\text{y}\sqrt{D})(x-\text{y}\sqrt{D})=1,$$

and,

$$\frac{1}{x+\text{y}\sqrt{D}}=x-\text{y}\sqrt{D},$$

is also an algebraic integer in $Q(\sqrt{D})$. The units of $Q(\sqrt{D})$ form an infinite multiplicative group (the group of Pell units). The question arises: What is the structure of this group?

A second property of rational numbers without an obvious analogue for algebraic numbers is related to the theorem on unique factorization of rational integers n in prime factors:

$$n=p_1^{a_1}...p_k^{a_k}.$$

For algebraic numbers such a unique factorization need not hold. E.g. consider the field $Q(\sqrt{-5})$. In it, the number 6 has two essentially different factorizations: $6 = 2.3, 6 = (1+\sqrt{-5})(1-\sqrt{-5})$. For extensions of higher degree the situation becomes still more complicated. The question arises: What becomes of the unique factorization theorem, does it have a meaning at all in algebraic number fields?

A third property without an obvious analogue is related to prime numbers. An ordinary prime number need not remain a prime number in an algebraic number field. E.g. in the field $Q(\sqrt{-1})$ of Gaussian numbers the prime number 5 has a factorization: $5 = (2+\sqrt{-1}+)(2-\sqrt{-1})$. However, the prime number 7 remains prime in this field. The question arises: Do there exist general laws governing the behaviour of prime numbers in algebraic number fields of higher degree? In other words, is it possible to find rules that would give a definite answer to the following question: Does a given prime number remain prime in a field K/Q or does it split in it, and if it splits, in how many factors?

Finally, a last (fourth) problem concerns the general structure of algebraic number fields. The field Q is the minimal field of characteristic zero without proper subfields. Any other algebraic number field does have subfields. E.g., Q is a subfield of any algebraic number field. The question arises: How many subfields are contained in a given extension K/Q, finitely or infinitely many, and what is their structure?

These are four main problems in algebraic number theory, and answering them constitutes the content of algebraic number theory. It is quite natural to start with the fourth problem, since its answer will shed light on the other three. The problem was solved by E. Galois in the 1820s. The fact that the number of subfields of an extension K/Q of degree n over Q is finite follows from the existence of a one-to-one correspondence (the fundamental Galois correspondence) between all subfields of the normal closure of K and the subgroups of its Galois group, a finite group of finite order (number of elements) at most $n!$.

The structure of the group of units of a field was elucidated by P. Dirichlet. His basic idea can be given by taking the group of Pell units as example. Any power of such a unit (both positive or negative) is a unit. There is a fundamental unit $\eta = x_0 + y_0\sqrt{D}$, and all other units are plus or minus integral powers of it; hence, the Pell units form the product of $\langle -1 \rangle$ with an infinite cyclic group. This fact is a special case of the general Dirichlet unit theorem for algebraic number fields: If the degree of a field K/Q is $n = r_1 + 2r_2$, where r_1 is the number of real imbeddings $K \to R$ and r_2 the number of complex-conjugate pairs of complex imbeddings $K \to C$, then the group of units U of K is the product of a finite cyclic group and $r = r_1 + r_2 - 1$ infinite cyclic groups:

$$U = C \times (\eta_1) \times \ldots \times (\eta_\tau).$$

Here $\eta_1, \ldots, \eta_\tau$ are independent fundamental units and C is the group of roots of unity contained in K. The norm of any unit in a field, i.e. the product of the unit by its conjugates, is equal to ± 1.

The problem of the non-unique factorization of algebraic integers in algebraic number fields was solved by E. Kummer, who started from a special case; he tried to solve Fermat's last theorem on the impossibility of solving the equation $x^p + y^p = z^p$ in non-zero integers for any prime number $p > 2$. Kummer expanded the left-hand side using $p-$th roots of unity, and hence the problem was reduced to one about the algebraic integers of $Q(\zeta), \zeta$ a primitive $p-$th root of unity. If there would be unique factorization into prime factors in $Q(\zeta)$, then it would have been sufficient to prove that not all prime factors at the left-hand side occur with an exponent that is a multiple of p.

This was Kummer's first point of view, but Dirichlet pointed out to him that unique factorization need not hold. In order to overcome this difficulty Kummer introduced ideal numbers, thus altering the entire structure of algebraic number theory for the future. The concept of an ideal number arises from the fact that if a field k does not contain prime numbers into which any algebraic integer in k can be split, then there is a field K / k of finite degree over k in which there does exist a collection of numbers that play the role of primes for k. These numbers were called ideal by Kummer (since they do not lie in the original field k). By introducing ideal numbers the theorem on unique factorization in k holds. Two numbers in a field that differ by a unit (so-called associated numbers) have one and the same ideal factors. (Note that ideal numbers are defined relative to the original field k;for another field k' one must construct an extension K'/k' (of possibly different degree over k') in which all ideal numbers of k' are contained).

Kummer also introduced the concept of the ideal class number of a field k. Two ideal numbers are said to belong to the same class if their quotient belongs to k. The number of classes thus obtained is called the ideal class number of k. He obtained the following important result: The class number h of k is finite, and the classes form an Abelian group under multiplication. Thus, any ideal number can be regarded as the $h-$th root of some element of the original field. The class number can be explicitly described in terms of other field constants (the regulator, the discriminant and the degree of the field).

Subsequently, the concept of an ideal number was replaced by that of an ideal, equivalent to it, which can be described in terms of the field k itself. Already in the 1950s the concept of an ideal was generalized to the more comprehensive concept of a divisor. Therefore, the modern theory of Kummer may be stated in terms of divisors. For algebraic number fields, however, the concepts of a divisor and an ideal coincide, and such fields only will be considered in the sequel. The concept of an ideal is related to that of non-associated numbers, which helps in understanding the deep connections between Kummer's theory and Dirichlet's theory of units. Although Kummer did not succeed in solving Fermat's problem, his ideas extended far beyond this problem and the concept of an ideal has now become fundamental in many branches of mathematics.

Since prime ideal numbers are defined relative to a field or, in modern terminology, since they are prime ideals in a field, the third problem on the factorization of ordinary prime numbers in an algebraic number field can be stated in the following general form. Suppose one is given a field k and a prime ideal p of its ring of algebraic integers. The question is: Does p remain prime in an extension $K = k(\theta)$. or does it split in a product of prime ideals of the ring of algebraic integers of K? In the latter case: By which law does it split? This question led to class field theory, a central part of modern algebraic number theory. The first solution of this problem was given by Kummer, who proved that if θ is a root of an irreducible polynomial $f(x)$ and $1, \theta, \ldots, \theta^{n-1}$ is an integral basis for O_K over O_K, then p splits in $k(\theta)$ "by the same law as fx" does in the residue field $(\bmod \mathrm{p})$. In other words, the factorization of p in $k(\theta)$ is determined by the congruence

$$f(x) \equiv f_1^{\alpha_1}(x) \ldots f_m^{\alpha_m}(x)(\bmod \mathrm{p}).$$

The corresponding factorization is called Kummer's formula (or Kummer's decomposition)

$$\mathrm{p} = \mathfrak{B}_1^{a_1} \ldots \mathfrak{B}_m^{\alpha_m},$$

where $\mathfrak{B}_i$ are prime ideals in $K = k(\theta)$.

In principle, this equation solves the third problem in algebraic number theory, but it is a local equation in the sense that it is necessary to check it for each prime ideal separately. The problem of partitioning the set of all prime ideals in classes such that the factorization law is the same in any given class and such that, moreover, it is possible to give a simple description of these classes, is solved by class field theory for extensions K/k with Abelian Galois group $\mathrm{Gal}(K/k)$.

Equation $\mathrm{p} = \mathfrak{B}_1^{a_1} \ldots \mathfrak{B}_m^{\alpha_m}$, yields a preliminary concept of a class. Let n be the degree of the extension K/k and let f_i be the relative degree of the ideal $\mathfrak{B}_i$. Computing the relative norms N_K/k of both parts of equation $\mathrm{p} = \mathfrak{B}_1^{a_1} \ldots \mathfrak{B}_m^{\alpha_m}$ leads to $n = \alpha_1 f_1 + \ldots + \alpha_m f_m$, in which α_i and f_i are natural numbers. For n fixed equation $n = \alpha_1 f_1 + \ldots + \alpha_m f_m$, has a finite number of solutions, so that the set of all prime ideals of k can be partitioned into a finite number of classes and one can collect the prime ideals whose Kummer decomposition corresponds to one solution $(a_i f_i)$ of $n = \alpha_1 f_1 + \ldots + \alpha_m f_m$ into one class. The number of prime ideals with the property that some $\alpha_i > 2$, is finite, and they all are divisors of the discriminant $\mathfrak{B}$ of K/k Only infinite classes are of interest, so that classes for which $\alpha_i > 2$ can be ignored.

In order to simplify the exposition, the field K/k is considered to be normal from now on. In such fields the condition $f_1 = \ldots = f_m$ holds. Therefore, the set of all p not dividing $\mathfrak{B}$ is partitioned into $d(n)$ classes, where $d(n)$ denotes the number of divisors. The class with $m = n$ is of special interest, in it

$$f = \ldots = f = .$$

A prime ideal p in it has maximal number of prime divisors in K/k:

$$\mathrm{p} = \mathrm{p}_1 \cdots \mathrm{p}_n.$$

Such p are said to split completely (or to be totally split), and their class is called the principal class of k relative to K/k. It is a principal object of study in class field theory. In order to be able to define the principal class via equation $\mathrm{p} = \mathrm{p}_1 \ldots \mathrm{p}_n$ it is necessary to give a proof of the facts that prime ideals satisfying equation $\mathrm{p} = \mathfrak{B}_1^{a_1} \ldots \mathfrak{B}_m^{\alpha_m}$, do exist in k and that there are infinitely many such ideals. Therefore a basic problem in class field theory is to define the principal class in terms of the field k itself in such a way that its infinite nature would follow. This problem has been completely solved for Abelian extensions K/k.

In order to be able to understand the ideas of class field theory better, it is necessary to introduce the general concept of an ideal class group. Kummer's definition given above then corresponds to the modern concept of an absolute ideal class group. The general concept of an ideal class group which is used nowadays is due to H. Weber and T. Takagi.

Weber introduced the concept of the conductor of a class group. Let m be an integral ideal of k, let L_{m} be the subgroup of principal ideals (α) of k given by $\alpha \equiv 1(\bmod \mathrm{m})$ and let A_{m} be the subgroup of all ideals of k that are relatively prime with m. A subgroup H_{m} satisfying $L_{\mathrm{m}} \subseteq H_{\mathrm{m}} \subseteq A_{\mathrm{m}}$ can be regarded as a group of principal ideals, and a (generalized) class group $A_{\mathrm{m}} / H_{\mathrm{m}}$ is constructed in this way. For $\mathrm{m}=1$ and $H_1 = L_1$ one obtains Kummer's definition. In the general case H_{m} consists of progressions $(\bmod \mathrm{m})$ whose residue classes form a subgroup of the whole multiplicative group $(\bmod \mathrm{m})$. The order h_{m} of this generalized class group satisfies the estimates $1 \leq h_{\mathrm{m}} \leq h.\phi(\mathrm{m})$, where h is the order of the absolute class group and ϕ is Euler's phi-function. For different conductors m_1 and m_2 the class groups may be equivalent if H_{m_1} and H_{m_2} consists in fact of the same progressions $(\bmod f)$, where $f = \mathrm{l.c.m.}(\mathrm{m}_1, \mathrm{m}_2)$. If one agrees not to distinguish between equivalent class groups, then one obtains the concept of a class group according to Weber with a conductor f that is the greatest common divisor of all equivalent conductors. A class field according to Weber is a field K / k in which only the prime ideals in its principal class H_f, and only these, split completely. Dirichlet's theorem on prime ideals in progressions, which is valid in every field k, implies that there are infinitely many prime ideals. Weber has shown that in certain particular cases the Galois group $\mathrm{Gal}(K / k)$ of a class field and the class group A_f / H_f of k are isomorphic.

A new point of view on class field theory, which is still valid, originated with D. Hilbert. He understood that between all relatively Abelian extensions of a field k and all class fields for this field there must exist a one-to-one correspondence. This correspondence can be stated as follows. If for some conductor f one constructs the Weber class group, then there is only one normal extension K / k in which the prime ideals of Weber's principal class, and only these, will split completely. Moreover, the Galois group of K / k is isomorphic to the Weber class group and the discriminant D of K / k consists of the same prime ideals as the conductor f. The converse is also true: If an Abelian extension with Galois group $\mathrm{Gal}(K / k)$ is given, then there exists a method (subsequently explicitly formulated by Takagi) by which one can uniquely construct a principal class H_f such that the class group A_f / H_f is isomorphic to $\mathrm{Gal}(K / k)$, only the prime ideals of H_f split completely in K, and such that the conductor f has the same prime divisors as the discriminant D of K / k, The "duality" was stated by Hilbert in 1900 as a hypothesis (he proved it only in special cases). In its general form it was proved by Takagi.

The next important stage of development of class group theory is related to the name of E. Artin, who revealed the special role of the canonical isomorphism between the Galois group and the ideal class group. He proved that this isomorphism is given by the Frobenius automorphism σ_{p} of an Abelian extension K / k, defined as:

$$a^{N_{\mathrm{p}}} = \sigma_{\mathrm{p}} \alpha (\bmod \mathrm{p}).$$

Here $N_{\mathfrak{p}}$ is the absolute norm of the ideal $\mathfrak{p}$, a runs through all numbers of K and $\mathfrak{p}$ is a prime ideal of k. The automorphism $\sigma_{\mathfrak{p}}$ (nowadays denoted by:

$$\left(\frac{K/k}{\mathfrak{p}}\right)),$$

depends (in an appropriate group) only on the ideal class to which $\mathfrak{p}$ belongs. It is multiplicative:

$$\left(\frac{K/k}{\mathfrak{p}_1}\right)\left(\frac{K/k}{\mathfrak{p}_2}\right)=\left(\frac{K/k}{\mathfrak{p}_1\mathfrak{p}_2}\right),$$

where the symbol at the right-hand side is understood as an automorphism of the class to which $\mathfrak{p}1\mathfrak{p}2$ belongs. With this in mind, Artin introduced the symbol.

$$\left(\frac{K\ \ k}{}\right),$$

on the entire group A_f of ideals a of k that are relatively prime with the conductor f. It is called the Artin symbol and realizes an isomorphism between the class group A_f/H_f and the Galois group $\mathrm{Gal}(K/k)$, whose explicit from is expressed in Artin's reciprocity law:

$$\left(\frac{K/k}{a}\right)=1,$$

if and only if $a \in H_f$ (reciprocity as a correspondence between the groups $\mathrm{Gal}(K/k)$ and A_f/H_f). From this one can obtain the classical form of the reciprocity law in the language of Kummer's power reciprocity symbol (one has to consider the field $K=k(a^{1/n})$, cf. Reciprocity laws). This form, in turn, implies the reciprocity law for Hilbert's symbol. In all three forms the reciprocity law is regarded as an isomorphism of groups, and the symbols of Artin, Kummer and Hilbert are regarded as group elements realizing this isomorphism. However, each of them also has a numerical value, which is equal to some $n-\mathrm{th}$ root of unity. Therefore one can formulate the reciprocity law as follows. Given the value of $(\alpha/\beta)_n$, it is required to find the value of the symbol $(\alpha/\beta)_n$ reciprocal to it, i.e. to exhibit the explicit form of the function of (α/β) defined by:

$$\left(\frac{\alpha}{\beta}\right)\left(\frac{\beta}{\alpha}\right)^{-1}.$$

In this form the law first appeared in the work of C.F. Gauss for quadratic fields and in the work of Kummer for cyclotomic fields of prime degree. The study of the reciprocity law in this form was subsequently continued by C.G.J. Jacobi, F.G.M. Eisenstein, Hilbert, H. Hasse, and others, but they only obtained partial results. I.R. Shafarevich solved this problem in its general form in 1948, basing himself on the idea of establishing a connection between the arithmetic definition of the symbol:

$$\left(\frac{\alpha,\beta}{\mathfrak{p}}\right),$$

and the analytic definition of an Abelian differential α, d, β on a Riemann surface. He gave a construction of:

$$\left(\frac{\alpha,\beta}{\mathrm{p}}\right),$$

which precisely corresponds to the determination of the residue of α, d, β in the p -adic field. For this purpose he introduced E -functions, also called Shafarevich functions, in terms of which he also obtained an explicit form of the reciprocity law in the general case.

In the late 1920s Hasse introduced the idea of doing class field theory for one prime ideal of k at a time, and reformulated and proved many theorems for an Abelian extension k_p of a local field k_p. At first the idea was of secondary importance only (the results were obtained as a consequence of the local theory), but at the end of the 1930s C. Chevalley proved the extraordinary importance of the local point of view in the creation of class field theory. He introduced the concept of an idèle group and formulated general class field theory from the local point of view. Following this the local-global principle became established in class field theory. Subsequently it was extended and refined, and as a result Abelian class field theory took a structured and completed form. The problem of creating non-Abelian class field theory for normal extensions with non-Abelian Galois group remains.

The exposition above relates mainly to the qualitative aspects of algebraic number theory. In questions of quantitative estimation and methods algebraic number theory is intimately connected with analytic number theory. It is also based, to a large extent, on properties of ζ -functions and L -functions of algebraic fields.

ALGEBRAIC NUMBER FIELD

In mathematics, an algebraic number field (or simply number field) *F* is a finite degree (and hence algebraic) field extension of the field of rational numbers Q. Thus *F* is a field that contains Q and has finite dimension when considered as a vector space over Q.

The notion of algebraic number field relies on the concept of a field. A field consists of a set of elements together with two operations, namely addition, and multiplication, and some distributivity assumptions. A prominent example of a field is the field of rational numbers, commonly denoted Q, together with its usual operations of addition and multiplication.

Another notion needed to define algebraic number fields is vector spaces. To the extent needed here, vector spaces can be thought of as consisting of sequences (or tuples)

$$(x_1, x_2, \ldots)$$

whose entries are elements of a fixed field, such as the field Q. Any two such sequences can be added by adding the entries one per one. Furthermore, any sequence can be multiplied by a single element *c* of the fixed field. These two operations known as vector addition and scalar multiplication satisfy a number of properties that serve to define vector spaces abstractly. Vector spaces are allowed to be "infinite-dimensional", that is to say that the sequences constituting the vector spaces are of infinite length. If, however, the vector space consists of *finite* sequences

$$(x_1, x_2, \ldots, x_n),$$

the vector space is said to be of finite dimension, n.

An algebraic number field (or simply number field) is a finite-degree field extension of the field of rational numbers. Here degree means the dimension of the field as a vector space over Q.

Examples:

- The smallest and most basic number field is the field Q of rational numbers. Many properties of general number fields are modeled after the properties of Q.
- The Gaussian rationals, denoted Q(i) (read as "Q adjoined i"), form the first nontrivial example of a number field. Its elements are expressions of the form $a+bi$ where both a and b are rational numbers and i is the imaginary unit. Such expressions may be added, subtracted, and multiplied according to the usual rules of arithmetic and then simplified using the identity $i^2 = -1$.

Explicitly,

$$(a + bi) + (c + di) = (a + c) + (b + d)i,$$

$$(a + bi)(c + di) = (ac - bd) + (ad + bc)i.$$

Non-zero Gaussian rational numbers are invertible, which can be seen from the identity

$$(a+bi)\left(\frac{a}{a^2+b^2} - \frac{b}{a^2+b^2}i\right) = \frac{(a+bi)(a-bi)}{a^2+b^2} = 1.$$

It follows that the Gaussian rationals form a number field which is two-dimensional as a vector space over Q.

- More generally, for any square-free integer d, the quadratic field $Q(\sqrt{d})$ is a number field obtained by adjoining the square root of d to the field of rational numbers. Arithmetic operations in this field are defined in analogy with the case of gaussian rational numbers, $d = -1$.
- Cyclotomic field $Q(\zeta_n)$, $\zeta_n = \exp(2\pi i / n)$ is a number field obtained from Q by adjoining a primitive nth root of unity ζ_n. This field contains all complex nth roots of unity and its dimension over Q is equal to $\varphi(n)$, where φ is the Euler totient function.
- The real numbers, R, and the complex numbers, C, are fields which have infinite dimension as Q-vector spaces, hence, they are *not* number fields. This follows from the uncountability of R and C as sets, whereas every number field is necessarily countable.
- The set Q^2 of ordered pairs of rational numbers, with the entrywise addition and

multiplication is a two-dimensional commutative algebra over Q. However, it is not a field, since it has zero divisors:

$$(1, 0) \cdot (0, 1) = (1 \cdot 0, 0 \cdot 1) = (0, 0).$$

Algebraicity and Ring of Integers

Generally, in abstract algebra, a field extension F / E is algebraic if every element f of the bigger field F is the zero of a polynomial with coefficients $e_0, \ldots, e_m$ in E:

$$p(f) = e_m f^m + e_{m-1} f^{m-1} + \ldots + e_1 f + e_0 = 0.$$

It is a fact that every field extension of finite degree is algebraic (proof: for x in F simply consider $1, x, x^2, x^3, \ldots$, we get a linear dependence, i.e. a polynomial that x is a root of!) because of the finite degree. In particular this applies to algebraic number fields, so any element f of an algebraic number field F can be written as a zero of a polynomial with rational coefficients. Therefore, elements of F are also referred to as *algebraic numbers*. Given a polynomial p such that $p(f) = 0$, it can be arranged such that the leading coefficient e_m is one, by dividing all coefficients by it, if necessary. A polynomial with this property is known as a monic polynomial. In general it will have rational coefficients. If, however, its coefficients are actually all integers, f is called an *algebraic integer*. Any (usual) integer $z \in Z$ is an algebraic integer, as it is the zero of the linear monic polynomial: $p(t) = t - z$.

It can be shown that any algebraic integer that is also a rational number must actually be an integer, hence the name "algebraic integer". Again using abstract algebra, specifically the notion of a finitely generated module, it can be shown that the sum and the product of any two algebraic integers is still an algebraic integer. It follows that the algebraic integers in F form a ring denoted O_F called the ring of integers of F. It is a subring of (that is, a ring contained in) F. A field contains no zero divisors and this property is inherited by any subring. Therefore, the ring of integers of F is an integral domain. The field F is the field of fractions of the integral domain O_F. This way one can get back and forth between the algebraic number field F and its ring of integers O_F. Rings of algebraic integers have three distinctive properties: firstly, O_F is an integral domain that is integrally closed in its field of fractions F. Secondly, O_F is a Noetherian ring. Finally, every nonzero prime ideal of O_F is maximal or, equivalently, the Krull dimension of this ring is one. An abstract commutative ring with these three properties is called a *Dedekind ring* (or Dedekind domain), in honor of Richard Dedekind, who undertook a deep study of rings of algebraic integers.

Unique Factorization

For general Dedekind rings, in particular rings of integers, there is a unique factorization of ideals into a product of prime ideals. For example, the ideal (6) in $Z[\sqrt{-5}]$ factors into prime ideals as,

$$(6) = (2, 1 + \sqrt{-5})(2,1 - \sqrt{-5})(3, 1 + \sqrt{-5})(3, 1 - \sqrt{-5}).$$

However, unlike Z as the ring of integers of Q, the ring of integers of a proper extension of Q need not admit unique factorization of numbers into a product of prime numbers or, more precisely, prime elements. This happens already for quadratic integers, for example in $O_{\mathbf{Q}(\sqrt{-5})} = Z[\sqrt{-5}]$, the uniqueness of the factorization fails:

$$6 = 2 \cdot 3 = (1 + \sqrt{-5}) \cdot (1 - \sqrt{-5}).$$

Using the norm it can be shown that these two factorization are actually inequivalent in the sense that the factors do not just differ by a unit in $O_{\mathbf{Q}(\sqrt{-5})}$ - Euclidean domains are unique factorization domains; for example Z[i], the ring of Gaussian integers, and Z[ω], the ring of Eisenstein integers, where ω is a cube root of unity (unequal to 1), have this property.

ζ-functions, *L*-functions and Class Number Formula

The failure of unique factorization is measured by the class number, commonly denoted h, the cardinality of the so-called ideal class group. This group is always finite. The ring of integers O_F possesses unique factorization if and only if it is a principal ring or, equivalently, if F has class number 1. Given a number field, the class number is often difficult to compute. The class number problem, going back to Gauss, is concerned with the existence of imaginary quadratic number fields (i.e., $Q\left(\sqrt{-d}\right)$, $d \geq 1$) with prescribed class number. The class number formula relates h to other fundamental invariants of F. It involves the Dedekind zeta function $\zeta_F(s)$, a function in a complex variable s, defined by:

$$\zeta_F(s) := \prod_{\mathfrak{p}} \frac{1}{1 - N(\mathfrak{p})^{-s}}.$$

(The product is over all prime ideals of O_F, $N(\mathfrak{p})$ denotes the norm of the prime ideal or, equivalently, the (finite) number of elements in the residue field $O_F / \mathfrak{p}$. The infinite product converges only for Re(s) > 1, in general analytic continuation and the functional equation for the zeta-function are needed to define the function for all s). The Dedekind zeta-function generalizes the Riemann zeta-function in that $\zeta_{\mathbf{Q}}(s) = \zeta(s)$.

The class number formula states that $\zeta_F(s)$ has a simple pole at $s = 1$ and at this point the residue is given by:

$$\frac{2^{r_1} \cdot (2\pi)^{r_2} \cdot h \cdot \text{Reg}}{w \cdot \sqrt{|D|}}.$$

Here r_1 and r_2 classically denote the number of real embeddings and pairs of complex embeddings of F, respectively. Moreover, Reg is the regulator of F, w the number of roots of unity in F and D is the discriminant of F.

Dirichlet L-functions $L(\chi, s)$ are a more refined variant of $\zeta(s)$. Both types of functions encode the arithmetic behavior of Q and F, respectively. For example, Dirichlet's theorem asserts that in any arithmetic progression

$$a, a + m, a + 2m, \ldots$$

with coprime a and m, there are infinitely many prime numbers. This theorem is implied by the fact that the Dirichlet L-function is nonzero at $s = 1$. Using much more advanced techniques including algebraic K-theory and Tamagawa measures, modern number theory deals with a description, if largely conjectural, of values of more general L-functions.

Bases for Number Fields

Integral Basis

An *integral basis* for a number field F of degree n is a set,

$$B = \{b_1, ..., b_n\}$$

of n algebraic integers in F such that every element of the ring of integers O_F of F can be written uniquely as a Z-linear combination of elements of B; that is, for any x in O_F we have,

$$x = m_1b_1 + ... + m_nb_n,$$

where the m_i are (ordinary) integers. It is then also the case that any element of F can be written uniquely as,

$$m_1b_1 + ... + m_nb_n,$$

where now the m_i are rational numbers. The algebraic integers of F are then precisely those elements of F where the m_i are all integers.

Working locally and using tools such as the Frobenius map, it is always possible to explicitly compute such a basis, and it is now standard for computer algebra systems to have built-in programs to do this.

Power Basis

Let F be a number field of degree n. Among all possible bases of F (seen as a Q-vector space), there are particular ones known as power bases, that are bases of the form,

$$B_x = \{1, x, x^2, ..., x^{n-1}\}$$

for some element $x \in F$. By the primitive element theorem, there exists such an x, called a primitive element. If x can be chosen in O_F and such that B_x is a basis of O_F as a free Z-module, then B_x is called a power integral basis, and the field F is called a monogenic field. An example of a number field that is not monogenic was first given by Dedekind. His example is the field obtained by adjoining a root of the polynomial $x^3 - x^2 - 2x - 8$.

Regular representation, trace and determinant.

Using the multiplication in F, the elements of the field F may be represented by n-by-n matrices

$$A = A(x) = (a_{ij})_{1 \le i,j \le n},$$

by requiring,

$$xe_i = \sum_{j=1}^{n} a_{ij}e_j, \quad a_{ij} \in Q.$$

Here $e_1, ..., e_n$ is a fixed basis for F, viewed as a Q-vector space. The rational numbers a_{ij} are uniquely determined by x and the choice of a basis since any element of F can be uniquely represented as a linear combination of the basis elements. This way of associating a matrix to any element of the field F is called the *regular representation*. The square matrix A represents the effect of multiplication by x in the given basis. It follows that if the element y of F is represented by a matrix B, then the product xy is represented by the matrix product BA. Invariants of matrices, such as the trace, determinant, and characteristic polynomial, depend solely on the field element x and not on the basis. In particular, the trace of the matrix $A(x)$ is called the *trace* of the field element x and denoted $\mathrm{Tr}(x)$, and the determinant is called the *norm* of x and denoted $\mathrm{N}(x)$.

By definition, standard properties of traces and determinants of matrices carry over to Tr and N: $\mathrm{Tr}(x)$ is a linear function of x, as expressed by $\mathrm{Tr}(x + y) = \mathrm{Tr}(x) + \mathrm{Tr}(y)$, $\mathrm{Tr}(\lambda x) = \lambda\,\mathrm{Tr}(x)$, and the norm is a multiplicative homogeneous function of degree n: $\mathrm{N}(xy) = \mathrm{N}(x)\,\mathrm{N}(y)$, $\mathrm{N}(\lambda x) = \lambda^n\,\mathrm{N}(x)$. Here λ is a rational number, and x, y are any two elements of F.

The *trace form* derives is a bilinear form defined by means of the trace, as $\mathrm{Tr}(x\,y)$. The *integral trace form*, an integer-valued symmetric matrix is defined as $t_{ij} = \mathrm{Tr}(b_i b_j)$, where $b_1, ..., b_n$ is an integral basis for F. The *discriminant* of F is defined as $\det(t)$. It is an integer, and is an invariant property of the field F, not depending on the choice of integral basis.

The matrix associated to an element x of F can also be used to give other, equivalent descriptions of algebraic integers. An element x of F is an algebraic integer if and only if the characteristic polynomial p_A of the matrix A associated to x is a monic polynomial with integer coefficients. Suppose that the matrix A that represents an element x has integer entries in some basis e. By the Cayley–Hamilton theorem, $p_A(A) = 0$, and it follows that $p_A(x) = 0$, so that x is an algebraic integer. Conversely, if x is an element of F which is a root of a monic polynomial with integer coefficients then the same property holds for the corresponding matrix A. In this case it can be proven that A is an integer matrix in a suitable basis of F. The property of being an algebraic integer is *defined* in a way that is independent of a choice of a basis in F.

Example:

Consider $F = \mathrm{Q}(x)$, where x satisfies $x^3 - 11x^2 + x + 1 = 0$. Then an integral basis is $[1, x, 1/2(x^2 + 1)]$, and the corresponding integral trace form is

$$\begin{bmatrix} 3 & 11 & 61 \\ 11 & 119 & 653 \\ 61 & 653 & 3589 \end{bmatrix}.$$

The "3" in the upper left hand corner of this matrix is the trace of the matrix of the map defined by the first basis element (1) in the regular representation of F on F. This basis element induces the identity map on the 3-dimensional vector space, F. The trace of the matrix of the identity map on a 3-dimensional vector space is 3.

The determinant of this is $1304 = 2^3 \cdot 163$, the field discriminant; in comparison the root discriminant, or discriminant of the polynomial, is $5216 = 2^5 \cdot 163$.

Places

Mathematicians of the nineteenth century assumed that algebraic numbers were a type of complex number. This situation changed with the discovery of p-adic numbers by Hensel in 1897; and now it is standard to consider all of the various possible embeddings of a number field *F* into its various topological completions at once.

A *place* of a number field *F* is an equivalence class of absolute values on *F*. Essentially, an absolute value is a notion to measure the size of elements *f* of *F*. Two such absolute values are considered equivalent if they give rise to the same notion of smallness (or proximity). In general, they fall into three regimes. Firstly (and mostly irrelevant), the trivial absolute value $|\ |_0$, which takes the value 1 on all non-zero *f* in *F*. The second and third classes are Archimedean places and non-Archimedean (or ultrametric) places. The completion of *F* with respect to a place is given in both cases by taking Cauchy sequences in *F* and dividing out null sequences, that is, sequences $(x_n)_{n \in \mathbf{N}}$ such that $|x_n|$ tends to zero when *n* tends to infinity. This can be shown to be a field again, the so-called completion of *F* at the given place.

For $F = Q$, the following non-trivial norms occur (Ostrowski's theorem): the (usual) absolute value, which gives rise to the complete topological field of the real numbers R. On the other hand, for any prime number *p*, the *p*-adic absolute values is defined by

$$|q|_p = p^{-n}, \text{ where } q = p^n\, a/b \text{ and } a \text{ and } b \text{ are integers not divisible by } p.$$

In contrast to the usual absolute value, the *p*-adic norm gets *smaller* when *q* is multiplied by *p*, leading to quite different behavior of Q_p vis-à-vis R.

Archimedean Places

The standard notation r_1 and r_2 for the number of real and complex embeddings is used, respectively.

Calculating the archimedean places of *F* is done as follows: let *x* be a primitive element of *F*, with minimal polynomial *f* (over Q). Over R, *f* will generally no longer be irreducible, but its irreducible (real) factors are either of degree one or two. Since there are no repeated roots, there are no repeated factors. The roots *r* of factors of degree one are necessarily real, and replacing *x* by *r* gives an embedding of *F* into R; the number of such embeddings is equal to the number of real roots of *f*. Restricting the standard absolute value on R to *F* gives an archimedean absolute value on *F*; such an absolute value is also referred to as a *real place* of *F*. On the other hand, the roots of factors of degree two are pairs of conjugate complex numbers, which allows for two conjugate embeddings into C. Either one of this pair of embeddings can be used to define an absolute value on *F*, which is the same for both embeddings since they are conjugate. This absolute value is called a *complex place* of *F*.

If all roots of *f* above are real (respectively, complex) or, equivalently, any possible embedding $F \subset C$ is actually forced to be inside R (resp. C), *F* is called totally real (resp. totally complex).

Non-archimedean or Ultrametric Places

To find the non-Archimedean places, let again *f* and *x* be as above. In Q_p, *f* splits in factors of various degrees, none of which are repeated, and the degrees of which add up to *n*, the degree of *f*. For each of these *p*-adically irreducible factors *t*, we may suppose that *x* satisfies *t* and obtain an embedding of *F* into an algebraic extension of finite degree over Q_p. Such a local field behaves in many ways like a number field, and the p-adic numbers may similarly play the role of the rationals; in particular, we can define the norm and trace in exactly the same way, now giving functions mapping to Q_p. By using this p-adic norm map N_t for the place t, we may define an absolute value corresponding to a given p-adically irreducible factor t of degree m by $|\theta|_t = |\mathbf{N}_t(\theta)|_p^{1/m}$. Such an absolute value is called an ultrametric, non-Archimedean or p-adic place of F.

For any ultrametric place v we have that $|x|_v \leq 1$ for any x in O_F, since the minimal polynomial for x has integer factors, and hence its p-adic factorization has factors in Z_p. Consequently, the norm term (constant term) for each factor is a p-adic integer, and one of these is the integer used for defining the absolute value for v.

Prime Ideals in O_F

For an ultrametric place v, the subset of O_F defined by $|x|_v < 1$ is an ideal P of O_F. This relies on the ultrametricity of v: given x and y in P, then

$$|x + y|_v \leq \max(|x|_v, |y|_v) < 1.$$

Actually, P is even a prime ideal.

Conversely, given a prime ideal P of O_F, a discrete valuation can be defined by setting $v_P(x) = n$ where n is the biggest integer such that $x \in P^n$, the n-fold power of the ideal. This valuation can be turned into an ultrametric place. Under this correspondence, (equivalence classes) of ultrametric places of F correspond to prime ideals of O_F. For F = Q, this gives back Ostrowski's theorem: any prime ideal in Z (which is necessarily by a single prime number) corresponds to a non-Archimedean place and vice versa. However, for more general number fields, the situation becomes more involved.

Yet another, equivalent way of describing ultrametric places is by means of localizations of O_F. Given an ultrametric place v on a number field F, the corresponding localization is the subring T of F of all elements x such that $|x|_v \leq 1$. By the ultrametric property T is a ring. Moreover, it contains O_F. For every element x of F, at least one of x or x^{-1} is contained in T. Actually, since $F^\times/T^\times$ can be shown to be isomorphic to the integers, T is a discrete valuation ring, in particular a local ring. Actually, T is just the localization of O_F at the prime ideal P. Conversely, P is the maximal ideal of T.

Altogether, there is a three-way equivalence between ultrametric absolute values, prime ideals, and localizations on a number field.

Ramification

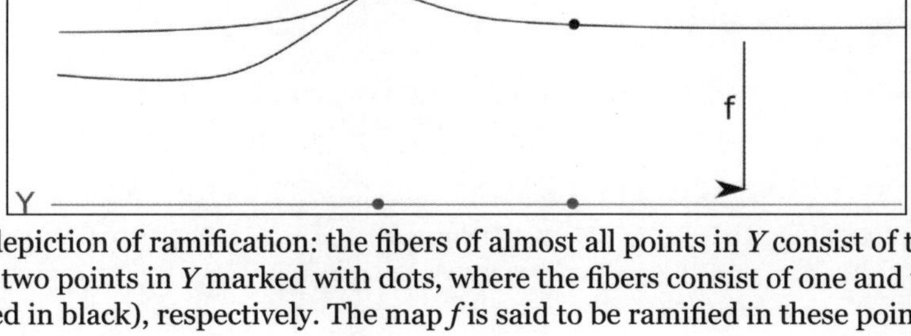

Schematic depiction of ramification: the fibers of almost all points in Y consist of three points, except for two points in Y marked with dots, where the fibers consist of one and two points (marked in black), respectively. The map f is said to be ramified in these points of Y.

Ramification, generally speaking, describes a geometric phenomenon that can occur with finite-to-one maps (that is, maps $f: X \to Y$ such that the preimages of all points y in Y consist only of finitely many points): the cardinality of the fibers $f^{-1}(y)$ will generally have the same number of points, but it occurs that, in special points y, this number drops. For example, the map

$$C \to C, z \mapsto z^n$$

has n points in each fiber over t, namely the n (complex) roots of t, except in $t = 0$, where the fiber consists of only one element, $z = 0$. One says that the map is "ramified" in zero. This is an example of a branched covering of Riemann surfaces. This intuition also serves to define ramification in algebraic number theory. Given a (necessarily finite) extension of number fields F / E, a prime ideal p of O_E generates the ideal pO_F of O_F. This ideal may or may not be a prime ideal, but, according to the Lasker–Noether theorem, always is given by:

$$pO_F = q_1^{e_1} q_2^{e_2} \dots q_m^{e_m}$$

with uniquely determined prime ideals q_i of O_F and numbers (called ramification indices) e_i. Whenever one ramification index is bigger than one, the prime p is said to ramify in F.

The connection between this definition and the geometric situation is delivered by the map of spectra of rings Spec $O_F \to$ Spec O_E. In fact, unramified morphisms of schemes in algebraic geometry are a direct generalization of unramified extensions of number fields.

Ramification is a purely local property, i.e., depends only on the completions around the primes p and q_i. The inertia group measures the difference between the local Galois groups at some place and the Galois groups of the involved finite residue fields.

An example:

The following example illustrates the notions introduced above. In order to compute the ramification index of Q(x), where $f(x) = x^3 - x - 1 = 0$, at 23, it suffices to consider the field extension $Q_{23}(x) / Q_{23}$. Up to $529 = 23^2$ (i.e., modulo 529) f can be factored as $f(x) = (x + 181)(x^2 - 181x - 38) = gh$.

Substituting $x = y + 10$ in the first factor g modulo 529 yields $y + 191$, so the valuation $|y|_g$ for y given by g is $|-191|_{23} = 1$. On the other hand, the same substitution in h yields $y^2 - 161y - 161$ modulo 529. Since $161 = 7 \times 23$,

$$|y|_h = \sqrt{|161|_{23}} = 1 / \sqrt{23}.$$

Since possible values for the absolute value of the place defined by the factor h are not confined to integer powers of 23, but instead are integer powers of the square root of 23, the ramification index of the field extension at 23 is two.

The valuations of any element of F can be computed in this way using resultants. If, for example $y = x^2 - x - 1$, using the resultant to eliminate x between this relationship and $f = x^3 - x - 1 = 0$ gives $y^{3-5}y^{2+4}y - 1 = 0$. If instead we eliminate with respect to the factors g and h of f, we obtain the corresponding factors for the polynomial for y, and then the 23-adic valuation applied to the constant (norm) term allows us to compute the valuations of y for g and h (which are both 1 in this instance).

Dedekind Discriminant Theorem

Much of the significance of the discriminant lies in the fact that ramified ultrametric places are all places obtained from factorizations in Q_p where p divides the discriminant. This is even true of the polynomial discriminant; however the converse is also true, that if a prime p divides the discriminant, then there is a p-place which ramifies. For this converse the field discriminant is needed. This is the Dedekind discriminant theorem. In the example above, the discriminant of the number field Q(x) with $x^3 - x - 1 = 0$ is −23, and as we have seen the 23-adic place ramifies. The Dedekind discriminant tells us it is the only ultrametric place which does. The other ramified place comes from the absolute value on the complex embedding of F.

Galois Groups and Galois Cohomology

Generally in abstract algebra, field extensions F / E can be studied by examining the Galois group Gal(F / E), consisting of field automorphisms of F leaving E elementwise fixed. As an example, the Galois group Gal (Q(ζ_n) / Q) of the cyclotomic field extension of degree n is given by $(Z/nZ)^\times$, the group of invertible elements in Z/nZ. This is the first stepstone into Iwasawa theory.

In order to include all possible extensions having certain properties, the Galois group concept is commonly applied to the (infinite) field extension $\overline{F}/F$ of the algebraic closure, leading to the absolute Galois group G := Gal($\overline{F}/F$) or just Gal(F), and to the extension F / Q. The fundamental theorem of Galois theory links fields in between F and its algebraic closure and closed subgroups of Gal (F). For example, the abelianization (the biggest abelian quotient) G^{ab} of G corresponds to a field referred to as the maximal abelian extension F^{ab} (called so since any further extension is not abelian, i.e., does not have an abelian Galois group). By the Kronecker–Weber theorem, the maximal abelian extension of Q is the extension generated by all roots of unity. For more general number fields, class field theory, specifically the Artin reciprocity law gives an answer by describing G^{ab} in terms of the idele class group. Also notable is the Hilbert class field, the maximal abelian unramified field extension of F. It can be shown to be finite over F, its Galois group over F is isomorphic to the class group of F, in particular its degree equals the class number h of F.

In certain situations, the Galois group acts on other mathematical objects, for example a group. Such a group is then also referred to as a Galois module. This enables the use of group cohomology for the Galois group Gal(F), also known as Galois cohomology, which in the first place measures the failure of exactness of taking Gal(F)-invariants, but offers deeper insights (and questions) as well. For example, the Galois group G of a field extension L / F acts on $L^\times$, the nonzero elements of L. This Galois module plays a significant role in many arithmetic dualities, such as Poitou-Tate duality. The Brauer group of F, originally conceived to classify division algebras over F, can be recast as a cohomology group, namely H^2(Gal (F), $\overline{F}^\times$).

Local-global Principle

Generally speaking, the term "local to global" refers to the idea that a global problem is first done at a local level, which tends to simplify the questions. Then, of course, the information gained in the local analysis has to be put together to get back to some global statement. For example, the notion of sheaves reifies that idea in topology and geometry.

Local and Global Fields

Number fields share a great deal of similarity with another class of fields much used in algebraic geometry known as function fields of algebraic curves over finite fields. An example is $F_p(T)$. They are similar in many respects, for example in that number rings are one-dimensional regular rings, as are the coordinate rings (the quotient fields of which is the function field in question) of curves. Therefore, both types of field are called global fields. In accordance with the philosophy laid out above, they can be studied at a local level first, that is to say, by looking at the corresponding local fields. For number fields F, the local fields are the completions of F at all places, including the archimedean ones. For function fields, the local fields are completions of the local rings at all points of the curve for function fields.

Many results valid for function fields also hold, at least if reformulated properly, for number fields. However, the study of number fields often poses difficulties and phenomena not encountered in function fields. For example, in function fields, there is no dichotomy into non-archimedean and archimedean places. Nonetheless, function fields often serves as a source of intuition what should be expected in the number field case.

Hasse Principle

A prototypical question, posed at a global level, is whether some polynomial equation has a solution in F. If this is the case, this solution is also a solution in all completions. The local-global principle or Hasse principle asserts that for quadratic equations, the converse holds, as well. Thereby, checking whether such an equation has a solution can be done on all the completions of F, which is often easier, since analytic methods (classical analytic tools such as intermediate value theorem at the archimedean places and p-adic analysis at the nonarchimedean places) can be used. This implication does not hold, however, for more general types of equations. However, the idea of passing from local data to global ones proves fruitful in class field theory, for example, where local class field theory is used to obtain global insights. This is also related to the fact that the Galois groups of the completions F_v can be explicitly determined, whereas the Galois groups of global fields, even of Q are far less understood.

Adeles and Ideles

In order to assemble local data pertaining to all local fields attached to F, the adele ring is set up. A multiplicative variant is referred to as ideles.

Gaussian Integer

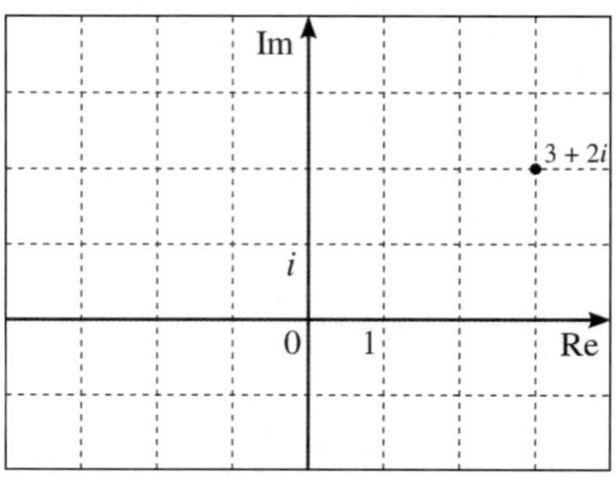

Gaussian integers as lattice points in the complex plane.

In number theory, a Gaussian integer is a complex number whose real and imaginary parts are both integers. The Gaussian integers, with ordinary addition and multiplication of complex numbers, form an integral domain, usually written as Z[*i*]. This integral domain is a particular case of a commutative ring of quadratic integers. It does not have a total ordering that respects arithmetic.

The Gaussian integers are the set:

$$Z[i] = \{a + bi \mid a, b \in Z\}, \qquad \text{where } i^2 = -1.$$

In other words, a Gaussian integer is a complex number such that its real and imaginary parts are both integers. Since the Gaussian integers are closed under addition and multiplication, they form a commutative ring, which is a subring of the field of complex numbers. It is thus an integral domain.

When considered within the complex plane, the Gaussian integers constitute the 2-dimensional integer lattice.

The *conjugate* of a Gaussian integer $a + bi$ is the Gaussian integer $a - bi$.

The *norm* of a Gaussian integer is its product with its conjugate.

$$N(a + bi) = (a + bi)(a - bi) = a^2 + b^2.$$

The norm of a Gaussian integer is thus the square of its absolute value as a complex number. The norm of a Gaussian integer is a nonnegative integer, which is a sum of two squares. Thus a norm cannot be of the form $4k + 3$, with k integer.

The norm is multiplicative, that is, one has,

$$N(zw) = N(z)N(w),$$

for every pair of Gaussian integers z, w. This can be shown directly, or by using the multiplicative property of the modulus of complex numbers.

The units of the ring of Gaussian integers (that is the Gaussian integers whose multiplicative inverse is also a Gaussian integer) are precisely the Gaussian integers with norm 1, that is, 1, –1, *i* and –*i*.

Euclidean Division

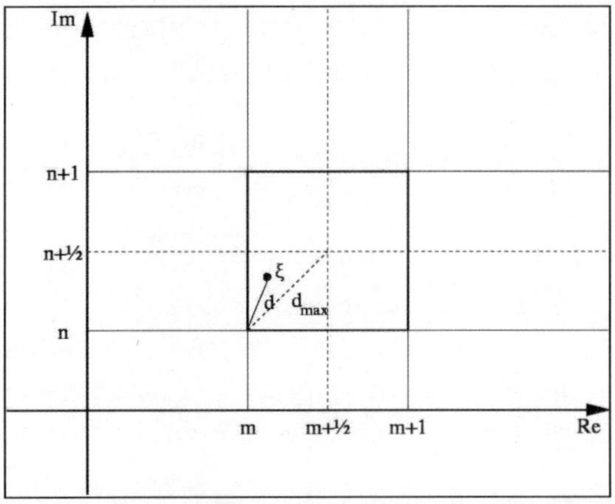

Visualization of maximal distance to some Gaussian integer.

Gaussian integers have a Euclidean division (division with remainder) similar to that of integers and polynomials. This makes the Gaussian integers a Euclidean domain, and implies that Gaussian integers share with integers and polynomials many important properties such as the existence of a Euclidean algorithm for computing greatest common divisors, Bézout's identity, the principal ideal property, Euclid's lemma, the unique factorization theorem, and the Chinese remainder theorem, all of which can be proved using only Euclidean division.

A Euclidean division algorithm takes, in the ring of Gaussian integers, a dividend a and divisor $b \neq 0$, and produces a quotient q and remainder r such that

$$a = bq + r \quad \text{and} \quad N(r) < N(b).$$

In fact, one may make the remainder smaller:

$$a = bq + r \quad \text{and} \quad N(r) \leq \frac{N(b)}{2}.$$

Even with this better inequality, the quotient and the remainder are not necessarily unique, but one may refine the choice to ensure uniqueness.

To prove this, one may consider the complex number quotient $x + iy = \frac{a}{b}$. There are unique integers m and n such that $-\frac{1}{2} < x - m \leq \frac{1}{2}$ and $\frac{1}{2} < y - n \leq \frac{1}{2}$, and thus $N(x - m + i(y - n)) \leq \frac{1}{2}$. Taking $q = m + in$, one has

$$a = bq + r,$$

with,

$$r = b(x - m + i(y - n)),$$

and,

$$N(r) \leq \frac{N(b)}{2}.$$

The choice of $x - m$ and $y - n$ in a semi-open interval is required for uniqueness. This definition of Euclidean division may be interpreted geometrically in the complex plane, by remarking that the distance from a complex number ξ to the closest Gaussian integer is at most $\frac{\sqrt{2}}{2}$.

Principal Ideals

Since the ring G of Gaussian integers is a Euclidean domain, G is a principal ideal domain, which means that every ideal of G is principal. Explicitly, an ideal I is a subset of a ring R such that every sum of elements of I and every product of an element of I by an element of R belong to I. An ideal is principal, if it consists of all multiples of a single element g, that is, it has the form $\{gx \mid x \in G\}$.

In this case, one says that the ideal is *generated* by g or that g is a *generator* of the ideal.

Every ideal I in the ring of the Gaussian integers is principal, because, if one chooses in I an non-zero element g of minimal norm, for every element x of I, the remainder of Euclidean division of x by g belongs also to I and has a norm that is smaller than that of g; because of the choice of g, this norm is zero, and thus the remainder is also zero. That is, one has $x = qg$, where q is the quotient.

For any g, the ideal generated by g is also generated by any *associate* of g, that is, $g, gi, -g, -gi$; no other element generates the same ideal. As all the generators of an ideal have the same norm, the *norm of an ideal* is the norm of any of its generators.

In some circumstances, it is useful to choose, once for all, a generator for each ideal. There are two classical ways for doing that, both considering first the ideals of odd norm. If the $g = a + bi$ has an odd norm $a^2 + b^2$, then one of a and b is odd, and the other is even. Thus g has exactly one associate with a real part a that is odd and positive. In his original paper, Gauss made another choice, by choosing the unique associate such that the remainder of its division by $2 + 2i$ is one. In fact, as $N(2 + 2i) = 8$, the norm of the remainder is not greater than 4. As this norm is odd, and 3 is not the norm of a Gaussian integer, the norm of the remainder is one, that is, the remainder is a unit. Multiplying g by the inverse of this unit, one finds an associate that has one as a remainder, when divided by $2 + 2i$.

If the norm of g is even, then either $g = 2^k h$ or $g = 2^k h(1+i)$, where k is a positive integer, and $N(h)$ is odd. Thus, one chooses the associate of g for getting a h which fits the choice of the associates for elements of odd norm.

Gaussian Primes

As the Gaussian integers form a principal ideal domain they form also a unique factorization domain. This implies that a Gaussian integer is irreducible (that is, it is not the product of two non-units) if and only if it is prime (that is, it generates a prime ideal).

The prime elements of Z[i] are also known as Gaussian primes. An associate of a Gaussian prime is also a Gaussian prime. The conjugate of a Gaussian prime is also a Gaussian prime (this implies that Gaussian primes are symmetric about the real and imaginary axes).

A positive integer is a Gaussian prime if and only if it is a prime number that is congruent to 3 modulo 4 (that is, it may be written $4n + 3$, with n a nonnegative integer) (sequence A002145 in the OEIS). The other prime numbers are not Gaussian primes, but each is the product of two conjugate Gaussian primes.

A Gaussian integer $a + bi$ is a Gaussian prime if and only if either:

- One of a, b is zero and absolute value of the other is a prime number of the form $4n + 3$ (with n a nonnegative integer), or
- Both are nonzero and $a^2 + b^2$ is a prime number (which will *not* be of the form $4n + 3$).

In other words, a Gaussian integer is a Gaussian prime if and only if either its norm is a prime number, or it is the product of a unit (± 1, $\pm i$) and a prime number of the form $4n + 3$.

It follows that there are three cases for the factorization of a prime number p in the Gaussian integers:

- If p is congruent to 3 modulo 4, then it is a Gaussian prime; in the language of algebraic number theory, p is said to be inert in the Gaussian integers.
- If p is congruent to 1 modulo 4, then it is the product of a Gaussian prime by its conjugate, both of which are non-associated Gaussian primes (neither is the product of the other by a unit); p is said to be a decomposed prime in the Gaussian integers. For example, $5 = (2 + i)(2 - i)$ and $13 = (3 + 2i)(3 - 2i)$.
- If $p = 2$, we have $2 = (1 + i)(1 - i) = i(1 - i)^2$; that is, 2 is the product of the square of a Gaussian prime by a unit; it is the unique ramified prime in the Gaussian integers.

Unique Factorization

As for every unique factorization domain, every Gaussian integer may be factored as a product of a unit and Gaussian primes, and this factorization is unique up to the order of the factors, and the replacement of any prime by any of its associates (together with a corresponding change of the unit factor).

If one chooses, once for all, a fixed Gaussian prime for each equivalence class of associated primes, and if one takes only these selected primes in the factorization, then one obtains a prime factorization which is unique up to the order of the factors. With the choices described above, the resulting unique factorization has the form

$$u(1+i)^{e_0} p_1^{e_1} \cdots p_k^{e_k},$$

where u is a unit (that is, $u \in \{1, -1, i, -i\}$), e_0 and k are nonnegative integers, $e_1, \ldots, e_k$ are positive integers, and $p_1, \ldots, p_k$ are distinct Gaussian primes such that, depending on the choice of selected associates,

- Either $p_k = a_k + ib_k$ with a odd and positive, and b even,
- Or the remainder of the euclidean division of p_k by $2 + 2i$ equals 1 (this is gauss's original choice).

An advantage of the second choice is that the selected associates behave well under products for Gaussian integers of odd norm. On the other hand, the selected associate for the real Gaussian primes are negative integers. For example, the factorization of 231 in the integers, and with the first choice of associates is $3 \times 7 \times 11$, while it is $(-1) \times (-3) \times (-7) \times (-11)$ with the second choice.

Gaussian Rationals

The field of Gaussian rationals is the field of fractions of the ring of Gaussian integers. It consists of the complex numbers whose real and imaginary part are both rational.

The ring of Gaussian integers is the integral closure of the integers in the Gaussian rationals.

This implies that Gaussian integers are quadratic integers and that a Gaussian rational is a Gaussian integer, if and only if it is a solution of an equation $x^2 + cx + d = 0$, with c and d integers. In fact $a + bi$ is solution of the equation $x^2 - 2ax + a^2 + b^2$, and this equation has integer coefficients if and only if a and b are both integers.

Greatest Common Divisor

As for any unique factorization domain, a *greatest common divisor (gcd)* of two Gaussian integers a, b is a Gaussian integer d that is a common divisor of a and b, which has all common divisors of a and b as divisor. That is (where | denotes the divisibility relation),

- $d \mid a$ and $d \mid b$, and
- $c \mid a$ and $c \mid b$ implies $c \mid d$.

Thus, *greatest* is meant relatively to the divisibility relation, and not for an ordering of the ring (for integers, both meanings of *greatest* coincide).

More technically, a greatest common divisor of a and b is a generator of the ideal generated by a and b (this characterization is valid for principal ideal domains, but not, in general, for unique factorization domains).

The greatest common divisor of two Gaussian integers is not unique, but is defined up to the multiplication by a unit. That is, given a greatest common divisor d of a and b, the greatest common divisors of a and b are d, $-d$, id, and $-id$.

There are several ways for computing a greatest common divisor of two Gaussian integers a and b. When one know prime factorizations of a and b,

$$a = i^k \prod_m p_m^{\nu_m}, \quad b = i^n \prod_m p_m^{\mu_m},$$

where the primes p_m are pairwise non associated, and the exponents μ_m non-associated, a greatest common divisor is $\prod_m p_m^{\lambda_m}$, with $\lambda_m = \min(\nu_m, \mu_m)$.

Unfortunately, except in simple cases, the prime factorization is difficult to compute, and Euclidean algorithm leads to a much easier (and faster) computation. This algorithm consists of replacing of the input (a, b) by (b, r), where r is the remainder of the Euclidean division of a by b, and repeating this operation until getting a zero remainder, that is a pair $(d, 0)$. This process terminates, because, at each step, the norm of the second Gaussian integer decreases. The resulting d is a greatest common divisor, because (at each step) b and $r = a - bq$ have the same divisors as a and b, and thus the same greatest common divisor.

This method of computation works always, but is not as simple as for integers because Euclidean division is more complicated. Therefore, a third method is often preferred for hand-written computations. It consists in remarking that the norm $N(d)$ of the greatest common divisor of a and b is a common divisor of $N(a)$, $N(b)$, and $N(a + b)$. When the greatest common divisor D of these three integers has few factors, then it is easy to test, for common divisor, all Gaussian integers with a norm dividing D.

For example, if $a = 5 + 3i$, and $b = 2 - 8i$, one has $N(a) = 34$, $N(b) = 68$, and $N(a + b) = 74$. As the greatest common divisor of the three norms is 2, the greatest common divisor of a and b has 1 or 2 as a norm. As a gaussian integer of norm 2 is necessary associated to $1 + i$, and as $1 + i$ divides a and b, then the greatest common divisor is $1 + i$.

If b is replaced by its conjugate $b = 2 + 8i$, then the greatest common divisor of the three norms is 34, the norm of a, thus one may guess that the greatest common divisor is a, that is, that $a \mid b$. In fact, one has $2 + 8i = (5 + 3i)(1 + i)$.

Congruence and Residue Classes

Given a Gaussian integer z_0 called a *modulus*, two Gaussian integers z_1, z_2 are *congruent modulo* z_0, if their difference is a multiple of z_0, that is if there exists a Gaussian integer q such that $z_1 - z_2 = qz_0$. In other words, two Gaussian integers are congruent modulo z_0, if their difference belongs to the ideal generated by z_0. This is denoted as $z_1 \equiv z_2 \ (mod\ z_0)$.

The congruence modulo z_0 is an equivalence relation (also called a congruence relation), which defines a partition of the Gaussian integers into equivalence classes, called here congruence classes or *residue classes*. The set of the residue classes is usually denoted $Z[i]/z_0Z[i]$, or $Z[i]/\langle z_0\rangle$, or simply $Z[i]/z_0$.

The residue class of a Gaussian integer a is the set

$$\bar{a} := \{z \in Z[i] \mid z \equiv a \pmod{z_0}\}$$

of all Gaussian integers that are congruent to a. It follows that $\bar{a} = \bar{b}$ if and only if $a \equiv b \ (\text{mod } z_0)$.

Addition and multiplication are compatible with congruences. This means that $a_1 \equiv b_1 \ (\text{mod } z_0)$ and $a_2 \equiv b_2 \ (\text{mod } z_0)$ imply $a_1 + a_2 \equiv b_1 + b_2 \ (\text{mod } z_0)$ and $a_1a_2 \equiv b_1b_2 \ (\text{mod } z_0)$. This defines well-defined operations (that is independent of the choice of representatives) on the residue classes:

$$\bar{a} + \bar{b} := \overline{a+b} \quad \text{and} \quad \bar{a} \cdot \bar{b} := \overline{ab}.$$

With these operations, the residue classes form a commutative ring, the quotient ring of the Gaussian integers by the ideal generated by z_0, which is also traditionally called the *residue class ring modulo* z_0.

Examples:

- There are exactly two residue classes for the modulus $1 + i$, namely $\bar{0} = \{0, \ \pm 2, \ \pm 4, \ldots, \pm 1 \ \pm i, \ \pm 3 \ \pm i, \ldots\}$ (all multiples of $1 + i$), and

  $\bar{1} = \{\pm 1, \ \pm 3, \ \pm 5, \ldots, \ \pm i, \ \pm 2 \ \pm i, \ldots\}$, which form a checkerboard pattern in the complex plane. These two classes form thus a ring with two elements, which is, in fact, a field, the unique (up to an isomorphism) field with two elements, and may thus be identified with the integers modulo 2. These two classes may be considered as a generalization of the partition of integers into even and odd integers. Thus one may speak of *even* and *odd* Gaussian integers (Gauss divided further even Gaussian integers into *even*, that is divisible by 2, and *half-even*).

- For the modulus 2 there are four residue classes, namely $\bar{0}, \bar{1}, \bar{i}, \overline{1+i}$. These form a ring with four elements, in which $x = -x$ for every x. Thus this ring is not isomorphic with the ring of integers modulo 4, another ring with four elements. One has $\overline{1+i}^2 = \bar{0}$, and thus this ring is not the finite field with four elements, nor the direct product of two copies of the ring of integers modulo 2.

- For the modulus 2 + 2i = $(i - 1)^3$ there are eight residue classes, namely $\overline{0}, \overline{\pm 1}, \overline{\pm i}, \overline{1 \pm i}, \overline{2}$, whereof four contain only even Gaussian integers and four contain only odd Gaussian integers.

Describing Residue Classes

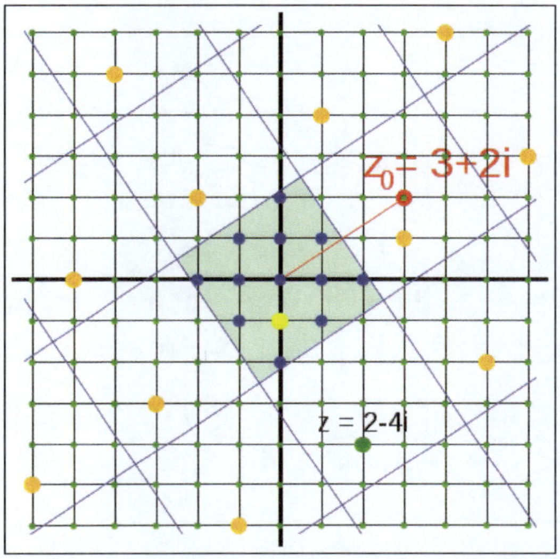

All 13 residue classes with their minimal residues (blue dots) in the square Q_{00} (light green background) for the modulus $z_0 = 3 + 2i$. One residue class with $z = 2 - 4i \equiv -i \pmod{z_0}$ is highlighted with yellow/orange dots.

Given a modulus z_0, all elements of a residue class have the same remainder for the Euclidean division by z_0, provided one uses the division with unique quotient and remainder, which is described above. Thus enumerating the residue classes is equivalent with enumerating the possible remainders. This can be done geometrically in the following way.

In the complex plane, one may consider a square grid, whose squares are delimited by the two lines

$$V_s = \left\{ z_0 \left(s - \tfrac{1}{2} + ix \right) \middle| x \in R \right\} \quad \text{and}$$

$$H_t = \left\{ z_0 \left(x + i \left(t - \tfrac{1}{2} \right) \right) \middle| x \in R \right\},$$

with s and t integers (blue lines in the figure). These divide the plane in semi-open squares (where m and n are integers)

$$Q_{mn} = \left\{ (s + it) z_0 \,\middle|\, s \in \left[m - \tfrac{1}{2}, m + \tfrac{1}{2} \right), t \in \left[n - \tfrac{1}{2}, n + \tfrac{1}{2} \right) \right\}.$$

The semi-open intervals that occur in the definition of Q_{mn} have been chosen in order that every complex number belong to exactly one square; that is, the squares Q_{mn} form a partition of the complex plane. One has

$$Q_{mn} = (m + in) z_0 + Q_{00} = \left\{ (m + in) z_0 + z \mid z \in Q_{00} \right\}.$$

This implies that every Gaussian integer is congruent modulo z_0 to a unique Gaussian integer in Q_{00} (the green square in the figure), which its remainder for the division by z_0. In other words, every residue class contains exactly one element in Q_{00}.

The Gaussian integers in Q_{00} (or in its boundary) are sometimes called *minimal residues* because their norm are not greater than the norm of any other Gaussian integer in the same residue class (Gauss called them *absolutely smallest residues*).

From this one can deduce by geometrical considerations, that the number of residue classes modulo a Gaussian integer $z_0 = a + bi$ equals his norm $N(z_0) = a^2 + b^2$.

Proof:

The relation $Q_{mn} = (m + in)z_0 + Q_{00}$ means that all Q_{mn} are obtained from Q_{00} by translating it by a Gaussian integer. This implies that all Q_{mn} have the same area $N = N(z_0)$, and contain the same number n_g of Gaussian integers.

Generally, the number of grid points (here the Gaussian integers) in an arbitrary square with the area A is $A+\Theta\left(\sqrt{A}\right)$. If one considers a big square consisting of $k \times k$ squares Q_{mn}, then it contains $k^2N+O(k\sqrt{N})$ grid points. It follows $k^2n_g = k^2N+\Theta(k\sqrt{N})$, and thus $n_g = N+\Theta(\frac{\sqrt{N}}{k})$, after a division by k^2. Taking the limit when k tends to the infinity gives $n_g = N = N(z_0)$.

Residue Class Fields

The residue class ring modulo a Gaussian integer z_0 is a field if and only if z_0 is a Gaussian prime.

If z_0 is a decomposed prime or the ramified prime $1 + i$ (that is, if its norm $N(z_0)$ is a prime number, which is either 2 or a prime congruent to 1 modulo 4), then the residue class field has a prime number of elements (that is, $N(z_0)$). It is thus isomorphic to the field of the integers modulo $N(z_0)$.

If, on the other hand, z_0 is an inert prime (that is, $N(z_0) = p^2$ is the square of a prime number, which is congruent to 3 modulo 4), then the residue class field has p^2 elements, and it is an extension of degree 2 (unique, up to an isomorphism) of the prime field with p elements (the integers modulo p).

Primitive Residue Class Group and Euler's Totient Function

Many theorems (and their proofs) for moduli of integers can be directly transferred to moduli of Gaussian integers, if one replaces the absolute value of the modulus by the norm. This holds especially for the *primitive residue class group* (also called multiplicative group of integers modulo n) and Euler's totient function. The primitive residue class group of a modulus z is defined as the subset of its residue classes, which contains all residue classes $\overline{a}$ that are coprime to z, i.e. $(a,z) = 1$. Obviously, this system builds a multiplicative group. The number of its elements shall be denoted by $\phi(z)$ (analogously to Euler's totient function $\varphi(n)$ for integers n).

For Gaussian primes it immediately follows that $\phi(p) = |p|^2 - 1$ and for arbitrary composite Gaussian integers,

$$z = i^k \prod_m p_m^{\nu_m}.$$

Euler's product formula can be derived as:

$$\phi(z) = \prod_{m(v_m>0)} |p_m^{v_m}|^2 \left(1 - \frac{1}{|p_m|^2}\right) = |z|^2 \prod_{p_m|z}\left(1 - \frac{1}{|p_m|^2}\right),$$

where the product is to build over all prime divisors p_m of z (with $v_m > 0$). Also the important theorem of Euler can be directly transferred:

For all a with $(a,z) = 1$, it holds that $a^{\phi(z)} \equiv 1 \pmod z$.

Quadratic Field

In algebraic number theory, a quadratic field is an algebraic number field K of degree two over Q, the rational numbers. The map $d \mapsto Q(\sqrt{d})$ is a bijection from the set of all square-free integers $d \neq 0, 1$ to the set of all quadratic fields. If $d > 0$ the corresponding quadratic field is called a real quadratic field, and for $d < 0$ an imaginary quadratic field or complex quadratic field, corresponding to whether it is or not a subfield of the field of the real numbers.

Quadratic fields have been studied in great depth, initially as part of the theory of binary quadratic forms. There remain some unsolved problems. The class number problem is particularly important.

Discriminant

For a nonzero square free integer d, the discriminant of the quadratic field $K = Q(\sqrt{d})$ is d if d is congruent to 1 modulo 4, and otherwise $4d$. For example, if d is -1, then K is the field of Gaussian rationals and the discriminant is -4. The reason for such a distinction is that the ring of integers of K is generated by $\frac{1}{2}\left(1+\sqrt{d}\right)$ in the first case, but by $\sqrt{d}$ in the second case.

The set of discriminants of quadratic fields is exactly the set of fundamental discriminants.

Prime Factorization into Ideals

Any prime number p gives rise to an ideal pO_K in the ring of integers O_K of a quadratic field K. In line with general theory of splitting of prime ideals in Galois extensions, this may be

p is inert:

(p) is a prime ideal.

The quotient ring is the finite field with p^2 elements: $O_K/pO_K = F_{p^2}$.

p splits:

(p) is a product of two distinct prime ideals of O_K.

The quotient ring is the product $O_K/pO_K = F_p \times F_p$.

p is ramified:

(p) is the square of a prime ideal of O_K.

The quotient ring contains non-zero nilpotent elements.

The third case happens if and only if p divides the discriminant D. The first and second cases occur when the Kronecker symbol (D/p) equals −1 and +1, respectively. For example, if p is an odd prime not dividing D, then p splits if and only if D is congruent to a square modulo p. The first two cases are in a certain sense equally likely to occur as p runs through the primes.

The law of quadratic reciprocity implies that the splitting behaviour of a prime p in a quadratic field depends only on p modulo D, where D is the field discriminant.

Quadratic Subfields of Cyclotomic Fields

The Quadratic Subfield of the Prime Cyclotomic Field

A classical example of the construction of a quadratic field is to take the unique quadratic field inside the cyclotomic field generated by a primitive p-th root of unity, with p a prime number > 2. The uniqueness is a consequence of Galois theory, there being a unique subgroup of index 2 in the Galois group over Q. As explained at Gaussian period, the discriminant of the quadratic field is p for $p = 4n + 1$ and $-p$ for $p = 4n + 3$. This can also be predicted from enough ramification theory. In fact p is the only prime that ramifies in the cyclotomic field, so that p is the only prime that can divide the quadratic field discriminant. That rules out the 'other' discriminants $-4p$ and $4p$ in the respective cases.

Other Cyclotomic Fields

If one takes the other cyclotomic fields, they have Galois groups with extra 2-torsion, and so contain at least three quadratic fields. In general a quadratic field of field discriminant D can be obtained as a subfield of a cyclotomic field of D-th roots of unity. This expresses the fact that the conductor of a quadratic field is the absolute value of its discriminant, a special case of the conductor-discriminant formula.

Orders of Quadratic Number Fields of Small Discriminant

The following table shows some orders of small discriminant of quadratic fields. The maximal order of an algebraic number field is its ring of integers, and the discriminant of the maximal order is the discriminant of the field. The discriminant of a non-maximal order is the product of the discriminant of the corresponding maximal order by the square of the determinant of the matrix that expresses a basis of the non-maximal order over a basis of the maximal order.

Order	Discriminant	Class number	Units	Comments
$Z[\sqrt{-5}]$	−20	2	±1	Ideal classes (1), $(1), (2, 1+\sqrt{-5})$
$Z[(1+\sqrt{-19})/2]$	−19	1	±1	Principal ideal domain, not Euclidean
$Z[2\sqrt{-1}]$	−16	1	±1	Non-maximal order
$Z[(1+\sqrt{-15})/2]$	−15	2	±1	Ideal classes (1), (2, $(1+\sqrt{-15})/2)$

$Z[\sqrt{-3}]$	−12	1	±1	Non-maximal order
$Z[(1+\sqrt{-11})/2]$	−11	1	±1	Euclidean
$Z[\sqrt{-2}]$	−8	1	±1	Euclidean
$Z[(1+\sqrt{-7})/2]$	−7	1	±1	Kleinian integers
$Z[\sqrt{-1}]$	−4	1	±1, ±i cyclic of order 4	Gaussian integers
$Z[(1+\sqrt{-3})/2]$	−3	1	$\pm 1, \left(\pm 1 \pm \sqrt{-3}\right)/2$	Eisenstein integers
$Z[\sqrt{-21}]$	-84	4		Class group non-cyclic $(C_2 \times C_2)$
$Z[(1+\sqrt{5})/2]$	5	1	$\pm((1+\sqrt{5})/2)^n$ (norm -1^n)	
$Z[\sqrt{2}]$	8	1	$\pm(1+\sqrt{2})^n$ (norm -1^n)	
$Z[\sqrt{3}]$	12	1	$\pm(2+\sqrt{3})^n$ (norm 1)	
$Z[(1+\sqrt{13})/2]$	13	1	$\pm((3+\sqrt{13})/2)^n$ (norm -1^n)	
$Z[(1+\sqrt{17})/2]$	17	1	$\pm(4+\sqrt{17})^n$ (norm -1^n)	
$Z[\sqrt{5}]$	20	2	$\pm(\sqrt{5}+2)^n$ (norm -1^n)	Non-maximal order

Cyclotomic Field

In number theory, a cyclotomic field is a number field obtained by adjoining a complex primitive root of unity to Q, the field of rational numbers. The n-th cyclotomic field $Q(\zeta_n)$ (where $n > 2$) is obtained by adjoining a primitive n-th root of unity ζ_n to the rational numbers.

The cyclotomic fields played a crucial role in the development of modern algebra and number theory because of their relation with Fermat's last theorem. It was in the process of his deep investigations of the arithmetic of these fields (for prime n) – and more precisely, because of the failure of unique factorization in their rings of integers – that Ernst Kummer first introduced the concept of an ideal number and proved his celebrated congruences.

Properties

A cyclotomic field is the splitting field of the cyclotomic polynomial

$$\Phi_n(x) = \prod_{\substack{1 \le k \le n \\ \gcd(k,n)=1}} \left(x - e^{2i\pi \frac{k}{n}} \right),$$

and therefore it is a Galois extension of the field of rational numbers. The degree of the extension

$$[Q(\zeta_n):Q],$$

is given by $\varphi(n)$ where φ is Euler's phi function. A complete set of Galois conjugates is given by $\{(\zeta_n)^a\}$, where a runs over the set of invertible residues modulo n (so that a is relative prime to n). The Galois group is naturally isomorphic to the multiplicative group

$$(Z/nZ)^\times$$

of invertible residues modulo n, and it acts on the primitive nth roots of unity by the formula

$$b\colon (\zeta_n)^a \to (\zeta_n)^{ab}.$$

The discriminant of the extension is

$$(-1)^{\varphi(n)/2} \frac{n^{\varphi(n)}}{\prod_{p|n} p^{\varphi(n)/(p-1)}},$$

where $\varphi(n)$ is Euler's totient function.

The ring of integers of the cyclotomic field $Q(\zeta_n)$ is $Z[\zeta_n]$.

Relation with Regular Polygons

Gauss made early inroads in the theory of cyclotomic fields, in connection with the geometrical problem of constructing a regular n-gon with a compass and straightedge. His surprising result that had escaped his predecessors was that a regular heptadecagon (with 17 sides) could be so constructed. More generally, if p is a prime number, then a regular p-gon can be constructed if and only if p is a Fermat prime; in other words if $\varphi(p) = p - 1 = 2^k$ is a power of 2.

For $n = 3$ and $n = 6$ primitive roots of unity admit a simple expression via square root of three, namely:

$$\zeta_3 = \frac{\sqrt{3}\,i - 1}{2}, \quad \zeta_6 = \frac{\sqrt{3}\,i + 1}{2}.$$

Hence, both corresponding cyclotomic fields are identical to the quadratic field $Q(\sqrt{-3})$. In the case of $\zeta_4 = i = \sqrt{-1}$ the identity of $Q(\zeta_4)$ to a quadratic field is even more obvious. However, this is not the case for $n = 5$, because expressing fifth roots of unity requires square roots of square root expressions, or a quadratic extension of a quadratic extension. The geometric problem for a general n can be reduced to the following question in Galois theory: can the nth cyclotomic field be built as a sequence of quadratic extensions?

Relation with Fermat's Last Theorem

A natural approach to proving Fermat's Last Theorem is to factor the binomial $x^n + y^n$, where n is an odd prime, appearing in one side of Fermat's equation $x^n + y^n = z^n$ as follows:

$$x^n + y^n = (x + y)(x + \zeta y)\cdots(x + \zeta^{n-1} y).$$

Here x and y are ordinary integers, whereas the factors are algebraic integers in the cyclotomic field $Q(\zeta_n)$. If unique factorization of algebraic integers were true, then it could have been used to rule out the existence of nontrivial solutions to Fermat's equation.

Several attempts to tackle Fermat's Last Theorem proceeded along these lines, and both Fermat's proof for $n = 4$ and Euler's proof for $n = 3$ can be recast in these terms. Unfortunately, the unique factorization fails in general – for example, for $n = 23$ – but Kummer found a way around this difficulty. He introduced a replacement for the prime numbers in the cyclotomic field $Q(\zeta_p)$, expressed the failure of unique factorization quantitatively via the class number h_p and proved that if h_p is not divisible by p (such numbers p are called *regular primes*) then Fermat's theorem is true for the exponent $n = p$. Furthermore, he gave a criterion to determine which primes are regular and using it, established Fermat's theorem for all prime exponents p less than 100, with the exception of the *irregular primes* 37, 59, and 67. Kummer's work on the congruences for the class numbers of cyclotomic fields was generalized in the twentieth century by Iwasawa in Iwasawa theory and by Kubota and Leopoldt in their theory of p-adic zeta functions.

Cubic Field

In mathematics, specifically the area of algebraic number theory, a cubic field is an algebraic number field of degree three.

If K is a field extension of the rational numbers Q of degree $[K:Q] = 3$, then K is called a cubic field. Any such field is isomorphic to a field of the form:

$$Q[x]/(f(x)),$$

where f is an irreducible cubic polynomial with coefficients in Q. If f has three real roots, then K is called a totally real cubic field and it is an example of a totally real field. If, on the other hand, f has a non-real root, then K is called a complex cubic field.

A cubic field K is called a cyclic cubic field, if it contains all three roots of its generating polynomial f. Equivalently, K is a cyclic cubic field if it is a Galois extension of Q, in which case its Galois group over Q is cyclic of order three. This can only happen if K is totally real. It is a rare occurrence in the sense that if the set of cubic fields is ordered by discriminant, then the proportion of cubic fields which are cyclic approaches zero as the bound on the discriminant approaches infinity.

A cubic field is called a pure cubic field, if it can be obtained by adjoining the real cube root $\sqrt[3]{n}$ of a cubefree positive integer n to the rational number field Q. Such fields are always complex cubic fields since each positive number has two complex non-real cube roots.

Examples:

- Adjoining the real cube root of 2 to the rational numbers gives the cubic field $Q(\sqrt[3]{2})$. This is an example of a pure cubic field, and hence of a complex cubic field. In fact, of all pure cubic fields, it has the smallest discriminant (in absolute value), namely −108.
- The complex cubic field obtained by adjoining to Q a root of $x^3 + x^2 - 1$ is not pure. It has the smallest discriminant (in absolute value) of all cubic fields, namely −23.
- Adjoining a root of $x^3 + x^2 - 2x - 1$ to Q yields a cyclic cubic field, and hence a totally real cubic field. It has the smallest discriminant of all totally real cubic fields, namely 49.

- The field obtained by adjoining to Q a root of $x^3 + x^2 - 3x - 1$ is an example of a totally real cubic field that is not cyclic. Its discriminant is 148, the smallest discriminant of a non-cyclic totally real cubic field.
- No cyclotomic fields are cubic because the degree of a cyclotomic field is equal to $\phi(n)$, where ϕ is Euler's totient function, which only takes on even values (except for $\phi(1) = \phi(2) = 1$).

Galois Closure

A cyclic cubic field K is its own Galois closure with Galois group Gal(K/Q) isomorphic to the cyclic group of order three. However, any other cubic field K is a non-galois extension of Q and has a field extension N of degree two as its Galois closure. The Galois group Gal(N/Q) is isomorphic to the symmetric group S_3 on three letters.

Associated Quadratic Field

The discriminant of a cubic field K can be written uniquely as df^2 where d is a fundamental discriminant. Then, K is cyclic if, and only if, $d = 1$, in which case the only subfield of K is Q itself. If $d \neq 1$, then the Galois closure N of K contains a unique quadratic field k whose discriminant is d (in the case $d = 1$, the subfield Q is sometimes considered as the "degenerate" quadratic field of discriminant 1). The conductor of N over k is f, and f^2 is the relative discriminant of N over K. The discriminant of N is d^3f^4.

The field K is a pure cubic field if, and only if, $d = -3$. This is the case for which the quadratic field contained in the Galois closure of K is the cyclotomic field of cube roots of unity.

Discriminant

Since the sign of the discriminant of a number field K is $(-1)^{r_2}$, where r_2 is the number of conjugate pairs of complex embeddings of K into **C**, the discriminant of a cubic field will be positive precisely when the field is totally real, and negative if it is a complex cubic field.

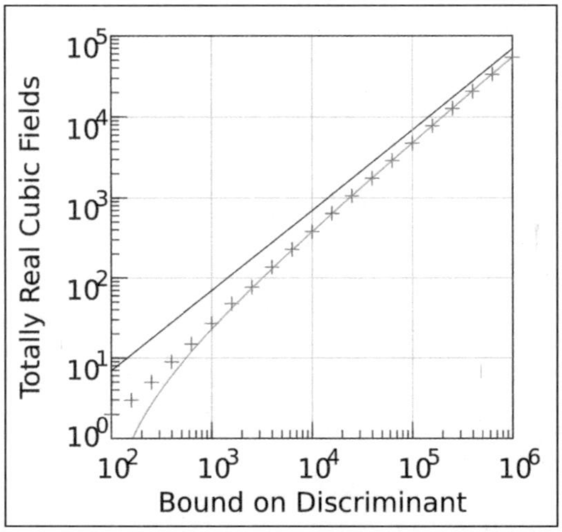

The blue crosses are the number of totally real cubic fields of bounded discriminant. The black line is the asymptotic distribution to first order whereas the green line includes the second order term.

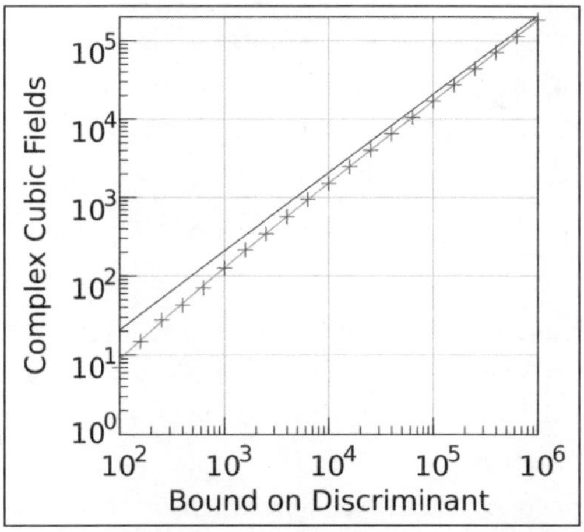

The blue crosses are the number of complex cubic fields of bounded discriminant. The black line is the asymptotic distribution to first order whereas the green line includes the second order term.

Given some real number $N > 0$ there are only finitely many cubic fields K whose discriminant D_K satisfies $|D_K| \leq N$. Formulae are known which calculate the prime decomposition of D_K, and so it can be explicitly calculated.

However, it should be pointed out that, different from quadratic fields, several non-isomorphic cubic fields $K_1, \ldots, K_m$ may share the same discriminant D. The number m of these fields is called the multiplicity of the discriminant D. Some small examples are $m = 2$ for $D = -1836, 3969$, $m = 3$ for $D = -1228, 22356$, $m = 4$ for $D = -3299, 32009$, and $m = 6$ for $D = -70956, 3054132$.

Any cubic field K will be of the form $K = Q(\theta)$ for some number θ that is a root of the irreducible polynomial

$$f(X) = X^3 - aX + b$$

with a and b both being integers. The discriminant of f is $\Delta = 4a^3 - 27b^2$. Denoting the discriminant of K by D, the index $i(\theta)$ of θ is then defined by $\Delta = i(\theta)^2 D$.

In the case of a non-cyclic cubic field K this index formula can be combined with the conductor formula $D = f^2 d$ to obtain a decomposition of the polynomial discriminant $\Delta = i(\theta)^2 f^2 d$ into the square of the product $i(\theta)f$ and the discriminant d of the quadratic field k associated with the cubic field K, where d is squarefree up to a possible factor 2^2 or 2^3. Georgy Voronoy gave a method for separating $i(\theta)$ and f in the square part of Δ.

The study of the number of cubic fields whose discriminant is less than a given bound is a current area of research. Let $N^+(X)$ (respectively $N^-(X)$) denote the number of totally real (respectively complex) cubic fields whose discriminant is bounded by X in absolute value. By means of an analysis of the residue of the Shintani zeta function, combined with a study of the tables of cubic fields compiled by Karim Belabas and some heuristics, David P. Roberts conjectured a more precise asymptotic formula:

$$N^{\pm}(X) \sim \frac{A_{\pm}}{12\zeta(3)} X + \frac{4\zeta(\frac{1}{3})B_{\pm}}{5\Gamma(\frac{2}{3})^3 \zeta(\frac{5}{3})} X^{\frac{5}{6}}.$$

where $A_\pm = 1$ or 3, $B_\pm = 1$ or $\sqrt{3}$, according to the totally real or complex case, $\zeta(s)$ is the Riemann zeta function, and $\Gamma(s)$ is the Gamma function. Proofs of this formula have been published by Bhargava, Shankar & Tsimerman using methods based on Bhargava's earlier work, as well as by Taniguchi & Thorne based on the Shintani zeta function.

Unit Group

According to Dirichlet's unit theorem, the torsionfree unit rank r of an algebraic number field K with r_1 real embeddings and r_2 pairs of conjugate complex embeddings is determined by the formula $r = r_1 + r_2 - 1$. Hence a totally real cubic field K with $r_1 = 3, r_2 = 0$ has two independent units ε_1, ε_2 and a complex cubic field K with $r_1 = r_2 = 1$ has a single fundamental unit ε_1. These fundamental systems of units can be calculated by means of generalized continued fraction algorithms by Voronoi, which have been interpreted geometrically by Delone and Faddeev.

Biquadratic Field

In mathematics, a biquadratic field is a number field K of a particular kind, which is a Galois extension of the rational number field Q with Galois group the Klein four-group.

Structure and Subfields

Biquadratic fields are all obtained by adjoining two square roots. Therefore in explicit terms they have the form

$$K = Q(\sqrt{a}, \sqrt{b}).$$

for rational numbers a and b. There is no loss of generality in taking a and b to be non-zero and square-free integers.

According to Galois theory, there must be three quadratic fields contained in K, since the Galois group has three subgroups of index 2. The third subfield, to add to the evident $Q(\sqrt{a})$ and $Q(\sqrt{b})$, is $Q(\sqrt{ab})$.

L-function

Biquadratic fields are the simplest examples of abelian extensions of Q that are not cyclic extensions. According to general theory the Dedekind zeta-function of such a field is a product of the Riemann zeta-function and three Dirichlet L-functions. Those L-functions are for the Dirichlet characters which are the Jacobi symbols attached to the three quadratic fields. Therefore taking the product of the Dedekind zeta-functions of the quadratic fields, multiplying them together, and dividing by the square of the Riemann zeta-function, is a recipe for the Dedekind zeta-function of the biquadratic field. This illustrates also some general principles on abelian extensions, such as the calculation of the conductor of a field.

Such L-functions have applications in analytic theory (Siegel zeroes), and in some of Kronecker's work.

QUADRATIC RECIPROCITY

In number theory, the law of quadratic reciprocity is a theorem about modular arithmetic that gives conditions for the solvability of quadratic equations modulo prime numbers. Due to its subtlety, it has many formulations, but the most standard statement is:

Law of quadratic reciprocity — Let p and q be distinct odd prime numbers, and define the Legendre symbol as:

$$\left(\frac{q}{p}\right)=\begin{cases}1 & \text{if } n^2\equiv q \pmod p \text{ for some integer } n,\\ -1 & \text{otherwise.}\end{cases}$$

Then:

$$\left(\frac{p}{q}\right)\left(\frac{q}{p}\right)=(-1)^{\frac{p-1}{2}\frac{q-1}{2}}.$$

This law allows the easy calculation of any Legendre symbol, making it possible to determine whether there is an integer solution for any quadratic equation of the form $x^2\equiv a \pmod p$ for p an odd prime; that is, to determine the "perfect squares" mod p. However, this is a non-constructive result: it gives no help at all for *finding* a specific solution; for this, one uses quadratic residues.

The theorem was conjectured by Euler and Legendre and first proved by Gauss. He refers to it as the "fundamental theorem" in the *Disquisitiones Arithmeticae* and his papers, writing

> The fundamental theorem must certainly be regarded as one of the most elegant of its type.

Privately he referred to it as the "golden theorem." He published six proofs, and two more were found in his posthumous papers. There are now over 240 published proofs.

Since Gauss, generalizing the reciprocity law has been a leading problem in mathematics, and has been crucial to the development of much of the machinery of modern algebra, number theory, and algebraic geometry, culminating in Artin reciprocity, class field theory, and the Langlands program.

Examples:

Quadratic reciprocity arises from certain subtle factorization patterns involving perfect square numbers.

Factoring $n^2 - 5$

Consider the polynomial $f(n)=n^2-5$ and its values for $n\in\mathbb{N}$. The prime factorizations of these values are given as follows:

n	$f(n)$			n	$f(n)$			n	$f(n)$	
1	−4	-2^2		16	251	251		31	956	$2^2\cdot 239$

2	-1	-1		17	284	$2^2\cdot 71$		32	1019	1019
3	4	2^2		18	319	$11\cdot 29$		33	1084	$2^2\cdot 271$
4	11	11		19	356	$2^2\cdot 89$		34	1151	1151
5	20	$2^2\cdot 5$		20	395	$5\cdot 79$		35	1220	$2^2\cdot 5\cdot 61$
6	31	31		21	436	$2^2\cdot 109$		36	1291	1291
7	44	$2^2\cdot 11$		22	479	479		37	1364	$2^2\cdot 11\cdot 31$
8	59	59		23	524	$2^2\cdot 131$		38	1439	1439
9	76	$2^2\cdot 19$		24	571	571		39	1516	$2^2\cdot 379$
10	95	$5\cdot 19$		25	620	$2^2\cdot 5\cdot 31$		40	1595	$5\cdot 11\cdot 29$
11	116	$2^2\cdot 29$		26	671	$11\cdot 61$		41	1676	$2^2\cdot 419$
12	139	139		27	724	$2^2\cdot 181$		42	1759	1759
13	164	$2^2\cdot 41$		28	779	$19\cdot 41$		43	1844	$2^2\cdot 461$
14	191	191		29	836	$2^2\cdot 11\cdot 19$		44	1931	1931
15	220	$2^2\cdot 5\cdot 11$		30	895	$5\cdot 179$		45	2020	$2^2\cdot 5\cdot 101$

The prime factors p dividing $f(n)$ are $p=2,5,$ and every prime whose final digit is 1or 9; no primes ending in 3 or 7 ever appear. Now, p is a prime factor of some n^2-5 whenever $n^2-5\equiv 0\,(mod\, p)$, i.e. whenever $n^2\equiv 5\,(mod\, p)$, i.e. whenever 5 is a quadratic residue modulo p. This happens for $p=2,5$ and those primes with $p\equiv 1,4\,(mod\ 5)$, and note that the latter numbers $1=(\pm 1)^2$ and $4=(\pm 2)^2$ are precisely the quadratic residues modulo 5. Therefore, except for $p=2,5$, we have that 5 is a quadratic residue modulo p iff p is a quadratic residue modulo 5.

The law of quadratic reciprocity gives a similar characterization of prime divisors of $f(n)=n^2-q$ for any prime q, which leads to a characterization for any integer q.

Patterns Among Quadratic Residues

Let p be an odd prime. A number modulo p is a quadratic residue whenever it is congruent to a square (mod p); otherwise it is a quadratic non-residue. ("Quadratic" can be dropped if it is clear from the context.) Here we exclude zero as a special case. Then as a consequence of the fact that the multiplicative group of a finite field of order p is cyclic of order p-1, the following statements hold:

- There are an equal number of quadratic residues and non-residues; and
- The product of two quadratic residues is a residue, the product of a residue and a non-residue is a non-residue, and the product of two non-residues is a residue.

For the avoidance of doubt, these statements do *not* hold if the modulus is not prime. For example, there are only 2 quadratic residues (1 and 4) in the multiplicative group modulo 15. Moreover although 7 and 8 are quadratic non-residues, their product 7 x 8 = 11 is also a quadratic non-residue, in contrast to the prime case.

The quadratic reciprocity law is the statement that certain patterns found in the table are true in general.

q = ±1 and the First Supplement

Trivially 1 is a quadratic residue for all primes. The question becomes more interesting for −1. Examining the table, we find −1 in rows 5, 13, 17, 29, 37, and 41 but not in rows 3, 7, 11, 19, 23, 31, 43 or 47. The former set of primes are all congruent to 1 modulo 4, and the latter are congruent to 3 modulo 4.

First Supplement to Quadratic Reciprocity: The congruence $x^2 \equiv -1(\bmod p)$ is solvable if and only if p is congruent to 1 modulo 4.

q = ±2 and the Second Supplement

Examining the table, we find 2 in rows 7, 17, 23, 31, 41, and 47, but not in rows 3, 5, 11, 13, 19, 29, 37, or 43. The former primes are all ≡ ±1 (mod 8), and the latter are all ≡ ±3 (mod 8). This leads to

Second Supplement to Quadratic Reciprocity: The congruence $x^2 \equiv 2\ (\bmod p)$ is solvable if and only if p is congruent to ±1 modulo 8.

−2 is in rows 3, 11, 17, 19, 41, 43, but not in rows 5, 7, 13, 23, 29, 31, 37, or 47. The former are ≡ 1 or ≡ 3 (mod 8), and the latter are ≡ 5, 7 (mod 8).

q = ±3

3 is in rows 11, 13, 23, 37, and 47, but not in rows 5, 7, 17, 19, 29, 31, 41, or 43. The former are ≡ ±1 (mod 12) and the latter are all ≡ ±5 (mod 12).

−3 is in rows 7, 13, 19, 31, 37, and 43 but not in rows 5, 11, 17, 23, 29, 41, or 47. The former are ≡ 1 (mod 3) and the latter ≡ 2 (mod 3).

Since the only residue (mod 3) is 1, we see that −3 is a quadratic residue modulo every prime which is a residue modulo 3.

q = ±5

5 is in rows 11, 19, 29, 31, and 41 but not in rows 3, 7, 13, 17, 23, 37, 43, or 47. The former are ≡ ±1 (mod 5) and the latter are ≡ ±2 (mod 5).

Since the only residues (mod 5) are ±1, we see that 5 is a quadratic residue modulo every prime which is a residue modulo 5.

−5 is in rows 3, 7, 23, 29, 41, 43, and 47 but not in rows 11, 13, 17, 19, 31, or 37. The former are ≡ 1, 3, 7, 9 (mod 20) and the latter are ≡ 11, 13, 17, 19 (mod 20).

Higher q

The observations about −3 and 5 continue to hold: −7 is a residue modulo p if and only if p is a residue modulo 7, −11 is a residue modulo p if and only if p is a residue modulo 11, 13 is a residue (mod p) if and only if p is a residue modulo 13, etc. The more complicated-looking rules for the quadratic characters of 3 and −5, which depend upon congruences modulo 12 and 20 respectively, are simply the ones for −3 and 5 working with the first supplement.

Example: For −5 to be a residue (mod p), either both 5 and −1 have to be residues (mod p) or they both have to be non-residues: i.e., $p \equiv \pm 1 \pmod 5$ *and* $p \equiv 1 \pmod 4$ or $p \equiv \pm 2 \pmod 5$ *and* $p \equiv 3 \pmod 4$. Using the Chinese remainder theorem these are equivalent to $p \equiv 1, 9 \pmod{20}$ or $p \equiv 3, 7 \pmod{20}$.

The generalization of the rules for −3 and 5 is Gauss's statement of quadratic reciprocity.

Legendre's Version

Another way to organize the data is to see which primes are residues mod which other primes, as illustrated in the following table. The entry in row p column q is R if q is a quadratic residue (mod p); if it is a nonresidue the entry is N.

If the row, or the column, or both, are $\equiv 1 \pmod 4$ the entry is blue or green; if both row and column are $\equiv 3 \pmod 4$, it is yellow or orange.

The blue and green entries are symmetric around the diagonal: The entry for row p, column q is R (resp N) if and only if the entry at row q, column p, is R (resp N).

The yellow and orange ones, on the other hand, are antisymmetric: The entry for row p, column q is R (resp N) if and only if the entry at row q, column p, is N (resp R).

The reciprocity law states that these patterns hold for all p and q.

Legend		
R	q is a residue (mod p)	q ≡ 1 (mod 4) or p ≡ 1 (mod 4) (or both)
N	q is a nonresidue (mod p)	
R	q is a residue (mod p)	both q ≡ 3 (mod 4) and p ≡ 3 (mod 4)
N	q is a nonresidue (mod p)	

Statement of the Theorem

Quadratic Reciprocity (Gauss's statement): If $q \equiv 1 \pmod 4$ then the congruence $x^2 \equiv p \pmod q$ is solvable if and only if $x^2 \equiv q \pmod p$ is solvable. If $q \equiv 3 \pmod 4$ then the congruence $x^2 \equiv p \pmod q$ is solvable if and only if $x^2 \equiv -q \pmod p$ is solvable.

Quadratic Reciprocity (combined statement): Define $q = (-\ )^{\overline{\ \ }}\ q$. Then the congruence $x^2 \equiv p \pmod q$ is solvable if and only if $x^2 \equiv q^* \pmod p$ is solvable.

Quadratic Reciprocity (Legendre's statement): If p or q are congruent to 1 modulo 4, then: $x^2 \equiv q \pmod p$ is solvable if and only if $x^2 \equiv p \pmod q$ is solvable. If p and q are congruent to 3 modulo 4, then: $x^2 \equiv q \pmod p$ is solvable if and only if $x^2 \equiv p \pmod q$ is not solvable.

The last is immediately equivalent to the modern form stated in the introduction above. It is a simple exercise to prove that Legendre's and Gauss's statements are equivalent – it requires no more than the first supplement and the facts about multiplying residues and nonresidues.

Alternative Statements

The theorem was formulated in many ways before its modern form: Euler and Legendre did not have Gauss's congruence notation, nor did Gauss have the Legendre symbol.

In this topic p and q always refer to distinct positive odd primes, and x and y to unspecified integers.

Fermat

Fermat proved (or claimed to have proved) a number of theorems about expressing a prime by a quadratic form:

$$p = x^2 + y^2 \quad \Leftrightarrow \quad p = 2 \quad \text{or} \quad p \equiv 1 \pmod 4$$
$$p = x^2 + 2y^2 \quad \Leftrightarrow \quad p = 2 \quad \text{or} \quad p \equiv 1,3 \pmod 8$$
$$p = x^2 + 3y^2 \quad \Leftrightarrow \quad p = 3 \quad \text{or} \quad p \equiv 1 \pmod 3$$

He did not state the law of quadratic reciprocity, although the cases -1, ± 2, and ± 3 are easy deductions from these and other of his theorems.

He also claimed to have a proof that if the prime number p ends with 7, (in base 10) and the prime number q ends in 3, and $p \equiv q \equiv 3 \pmod 4$, then

$$pq = x^2 + 5y^2.$$

Euler conjectured, and Lagrange proved, that

$$p \equiv 1,9 \pmod{20} \Rightarrow p = x^2 + 5y^2$$
$$p,q \equiv 3,7 \pmod{20} \Rightarrow pq = x^2 + 5y^2$$

Proving these and other statements of Fermat was one of the things that led mathematicians to the reciprocity theorem.

Euler

Translated into modern notation, Euler stated that for distinct odd primes p and q:

- If $q \equiv 1 \pmod 4$ then q is a quadratic residue (mod p) if and only if there exists some integer b such that $p \equiv b^2 \pmod q$.

- If $q \equiv 3 \pmod 4$ then q is a quadratic residue (mod p) if and only if there exists some integer b which is odd and not divisible by q such that $p \equiv \pm b^2 \pmod{4q}$.

This is equivalent to quadratic reciprocity.

He could not prove it, but he did prove the second supplement.

Legendre and his Symbol

Fermat proved that if p is a prime number and a is an integer,

$$a^p \equiv a(\operatorname{mod} p).$$

Thus if p does not divide a, using the non-obvious fact that the residues modulo p form a field and therefore in particular the multiplicative group is cyclic, hence there can be at most two solutions to a quadratic equation:

$$a^{\frac{p-1}{2}} \equiv \pm 1(\operatorname{mod} p).$$

Legendre lets a and A represent positive primes $\equiv 1 \pmod 4$ and b and B positive primes $\equiv 3 \pmod 4$, and sets out a table of eight theorems that together are equivalent to quadratic reciprocity:

Theorem	When	it follows that
I	$b^{\frac{a-1}{2}} \equiv 1 (\operatorname{mod} a)$	$a^{\frac{b-1}{2}} \equiv 1 (\operatorname{mod} b)$
II	$a^{\frac{b-1}{2}} \equiv -1 (\operatorname{mod} b)$	$b^{\frac{a-1}{2}} \equiv -1 (\operatorname{mod} a)$
III	$a^{\frac{A-1}{2}} \equiv 1 (\operatorname{mod} A)$	$A^{\frac{a-1}{2}} \equiv 1 (\operatorname{mod} a)$
IV	$a^{\frac{A-1}{2}} \equiv -1 (\operatorname{mod} A)$	$A^{\frac{a-1}{2}} \equiv -1 (\operatorname{mod} a)$
V	$a^{\frac{}{}} \equiv \quad (\operatorname{mod} b)$	$b^{\frac{a-1}{2}} \equiv 1 (\operatorname{mod} a)$
VI	$b^{\frac{a-1}{2}} \equiv -1 (\operatorname{mod} a)$	$a^{\frac{b-1}{2}} \equiv -1 (\operatorname{mod} b)$
VII	$b^{\frac{B-1}{2}} \equiv 1 (\operatorname{mod} B)$	$B^{\frac{b-1}{2}} \equiv -1 (\operatorname{mod} b)$
VIII	$b^{\frac{B-1}{2}} \equiv -1 (\operatorname{mod} B)$	$B^{\frac{b-1}{2}} \equiv 1 (\operatorname{mod} b)$

He says that since expressions of the form:

$$N^{\frac{c-1}{2}} (\operatorname{mod} c), \qquad \gcd(N,c) = 1$$

will come up so often he will abbreviate them as:

$$\left(\frac{N}{c}\right) \equiv N^{\frac{c-1}{2}} (\operatorname{mod} c) = \pm 1.$$

This is now known as the Legendre symbol, and an equivalent definition is used today: for all integers a and all odd primes p

$$\left(\frac{a}{p}\right)=\begin{cases}0 & a\equiv 0\pmod p\\ 1 & a\not\equiv 0\pmod p \text{ and } \exists x: a\equiv x^2\pmod p\\ -1 & a\not\equiv 0\pmod p \text{ and there is no such } x.\end{cases}$$

Legendre's version of quadratic reciprocity:

$$\left(\frac{p}{q}\right)=\begin{cases}\left(\frac{q}{p}\right) & p\equiv 1\pmod 4 \quad \text{or} \quad q\equiv 1\pmod 4\\ -\left(\frac{q}{p}\right) & p\equiv 3\pmod 4 \quad \text{and} \quad q\equiv 3\pmod 4\end{cases}$$

He notes that these can be combined:

$$\left(\frac{p}{q}\right)\left(\frac{q}{p}\right)=(-1)^{\frac{p-1}{2}\frac{q-1}{2}}.$$

A number of proofs, especially those based on Gauss's Lemma, explicitly calculate this formula.

The supplementary laws using Legendre symbols:

$$\left(\frac{-1}{p}\right)=(-1)^{\frac{p-1}{2}}=\begin{cases}1 & p\equiv 1\pmod 4\\ -1 & p\equiv 3\pmod 4\end{cases}$$

$$\left(\frac{2}{p}\right)=(-1)^{\frac{p^2-1}{8}}=\begin{cases}1 & p\equiv 1,7\pmod 8\\ -1 & p\equiv 3,5\pmod 8\end{cases}$$

Legendre's attempt to prove reciprocity is based on a theorem of his:

Legendre's Theorem: Let a, b and c be integers where any pair of the three are relatively prime. Moreover assume that at least one of ab, bc or ca is negative (i.e. they don't all have the same sign). If,

$$u^2\equiv -bc\pmod a$$
$$v^2\equiv -ca\pmod b$$
$$w^2\equiv -ab\pmod c$$

are solvable then the following equation has a nontrivial solution in integers:

$$ax^2+by^2+cz^2=0.$$

Example: Theorem is handled by letting $a \equiv 1$ and $b \equiv 3 \pmod 4$ be primes and assuming that $\left(\frac{b}{a}\right)=1$ and, contrary the theorem, that $\left(\frac{a}{b}\right)=-1$. Then $x^2+ay^2-bz^2=0$ has a solution, and taking congruences (mod 4) leads to a contradiction.

This technique doesn't work for Theorem VIII. Let $b \equiv B \equiv 3 \pmod 4$, and assume

$$\left(\frac{B}{b}\right)=\left(\frac{b}{B}\right)=-1.$$

Then if there is another prime $p \equiv 1 \pmod 4$ such that

$$\left(\frac{p}{b}\right)=\left(\frac{p}{B}\right)=-1,$$

the solvability of $Bx^2+by^2-pz^2=0$ leads to a contradiction (mod 4). But Legendre was unable to prove there has to be such a prime p; he was later able to show that all that is required is:

Legendre's Lemma: If p is a prime that is congruent to 1 modulo 4 then there exists an odd prime q such that $\left(\frac{p}{q}\right)=-1.$

but he couldn't prove that either. Hilbert symbol discusses how techniques based on the existence of solutions to $ax^2+by^2+cz^2=0$ can be made to work.

Gauss

Gauss first proves the supplementary laws. He sets the basis for induction by proving the theorem for ±3 and ±5. Noting that it is easier to state for −3 and +5 than it is for +3 or −5, he states the general theorem in the form:

If p is a prime of the form $4n + 1$ then p, but if p is of the form $4n + 3$ then $-p$, is a quadratic residue (resp. nonresidue) of every prime, which, with a positive sign, is a residue (resp. nonresidue) of p. In the next sentence, he christens it the "fundamental theorem" (Gauss never used the word "reciprocity").

Introducing the notation a R b (resp. a N b) to mean a is a quadratic residue (resp. nonresidue) (mod b), and letting a, a', etc. represent positive primes ≡ 1 (mod 4) and b, b', etc. positive primes ≡ 3 (mod 4), he breaks it out into the same 8 cases as Legendre:

Case	If	Then
1)	±a R a′	±a′ R a
2)	±a N a′	±a′ N a
3)	+a R b −a N b	±b R a
4)	+a N b −a R b	±b N a
5)	±b R a	+a R b−a N b
6)	±b N a	+a N b −a R b
7)	+b R b′−b N b′	−b′ N b+b′ R b
8)	−b N b′+b R b′	+b′ R b−b′ N b

Letting A, A', etc. represent any (prime or composite) positive numbers ≡ 1 (mod 4) and B, B', etc. positive numbers ≡ 3 (mod 4):

Case	If	Then
9)	±a R A	±A R a
10)	±b R A	+A R b−A N b
11)	+a R B	±B R a
12)	−a R B	±B N a
13)	+b R B	−B N b+N R b
14)	−b R B	+B R b−B N b

All of these cases take the form "if a prime is a residue (mod a composite), then the composite is a residue or nonresidue (mod the prime), depending on the congruences (mod 4)". He proves that these follow from cases 1) - 8).

Gauss needed, and was able to prove, a lemma similar to the one Legendre needed:

Gauss's Lemma: If p is a prime congruent to 1 modulo 8 then there exists an odd prime q such that:

$$q < 2\sqrt{p}+1 \quad and \quad \left(\frac{p}{q}\right) = -1.$$

The proof of quadratic reciprocity uses complete induction.

Gauss's Version in Legendre Symbols:

$$\left(\frac{p}{q}\right) = \begin{cases} \left(\frac{q}{p}\right) & q \equiv 1 \pmod 4 \\ \left(\frac{-q}{p}\right) & q \equiv 3 \pmod 4. \end{cases}$$

These can be combined:

Gauss's Combined Version in Legendre Symbols:

Let,

$$q^* = (-1)^{\frac{q-1}{2}} q.$$

In other words:

$$|q^*| = |q| \quad and \quad q^* \equiv 1 \pmod 4.$$

Then:

$$\left(\frac{p}{q}\right) = \left(\frac{q^*}{p}\right).$$

A number of proofs of the theorem, especially those based on Gauss sums derive this formula. or the splitting of primes in algebraic number fields.

Other Statements

Note that the statements are equivalent to quadratic reciprocity: if, for example, Euler's version is assumed, the Legendre-Gauss version can be deduced from it, and vice versa.

Euler's Formulation of Quadratic Reciprocity: If $p \equiv \pm q \pmod{4a}$ then $\left(\frac{a}{p}\right) = \left(\frac{a}{q}\right)$.

This can be proven using Gauss's lemma.

Quadratic Reciprocity (Gauss; Fourth Proof): Let $a, b, c, \ldots$ be unequal positive odd primes, whose product is n, and let m be the number of them that are $\equiv 3 \pmod 4$; check whether n/a is a residue of a, whether n/b is a residue of b, ... The number of nonresidues found will be even when $m \equiv 0, 1 \pmod 4$, and it will be odd if $m \equiv 2, 3 \pmod 4$.

Gauss's fourth proof consists of proving this theorem (by comparing two formulas for the value of Gauss sums) and then restricting it to two primes. He then gives an example: Let $a = 3$, $b = 5$, $c = 7$, and $d = 11$. Three of these, 3, 7, and $11 \equiv 3 \pmod 4$, so $m \equiv 3 \pmod 4$. $5\times7\times11$ R 3; $3\times7\times11$ R 5; $3\times5\times11$ R 7; and $3\times5\times7$ N 11, so there are an odd number of nonresidues.

Eisenstein's Formulation of Quadratic Reciprocity:

Assume,

$$p \neq q, \quad p' \neq q', \quad p \equiv p' \pmod 4, \quad q \equiv q' \pmod 4.$$

Then,

$$\left(\frac{p}{q}\right)\left(\frac{q}{p}\right) = \left(\frac{p'}{q'}\right)\left(\frac{q'}{p'}\right).$$

Mordell's Formulation of Quadratic Reciprocity: Let a, b and c be integers. For every prime, p, dividing abc if the congruence

$$ax^2 + by^2 + cz^2 \equiv 0 \left(\bmod \frac{4abc}{p}\right),$$

has a nontrivial solution, then so does:

$$ax^2 + by^2 + cz^2 \equiv 0 \pmod{4abc}.$$

Zeta function formulation: Quadratic reciprocity is equivalent to the zeta function of a quadratic field being the product of the Riemann zeta function and a certain Dirichlet L-function.

Jacobi Symbol

The Jacobi symbol is a generalization of the Legendre symbol; the main difference is that the bottom number has to be positive and odd, but does not have to be prime. If it is prime, the two symbols agree. It obeys the same rules of manipulation as the Legendre symbol. In particular

$$\left(\frac{-1}{n}\right) = (-1)^{\frac{n-1}{2}} = \begin{cases} 1 & n \equiv 1 \pmod 4 \\ -1 & n \equiv 3 \pmod 4 \end{cases}$$

$$\left(\frac{2}{n}\right)=(-1)^{\frac{n^2-1}{8}}=\begin{cases}1 & n\equiv 1,7\pmod 8\\ -1 & n\equiv 3,5\pmod 8),\end{cases}$$

and if both numbers are positive and odd (this is sometimes called "Jacobi's reciprocity law"):

$$\left(\frac{m}{n}\right)=(-1)^{\frac{(m-1)(n-1)}{4}}\left(\frac{n}{m}\right).$$

However, if the Jacobi symbol is 1 but the denominator is not a prime, it does not necessarily follow that the numerator is a quadratic residue of the denominator. Gauss's cases 9) - 14) above can be expressed in terms of Jacobi symbols:

$$\left(\frac{M}{p}\right)=(-1)^{\frac{(p-1)(M-1)}{4}}\left(\frac{p}{M}\right),$$

and since p is prime the left hand side is a Legendre symbol, and we know whether M is a residue modulo p or not.

The formulas listed are true for Jacobi symbols as long as the symbols are defined. Euler's formula may be written:

$$\left(\frac{a}{m}\right)=\left(\frac{a}{m\pm 4an}\right),\qquad n\in\mathbb{Z}, m\pm 4an>0.$$

Example:

$$\left(\frac{2}{7}\right)=\left(\frac{2}{15}\right)=\left(\frac{2}{23}\right)=\left(\frac{2}{31}\right)=\cdots=1.$$

2 is a residue modulo the primes 7, 23 and 31:

$$3^2\equiv 2\pmod 7),\quad 5^2\equiv 2\pmod{23}),\quad 8^2\equiv 2\pmod{31}).$$

But 2 is not a quadratic residue modulo 5, so it can't be one modulo 15. This is related to the problem Legendre had: if $\left(\frac{a}{m}\right)=-1$, then a is a non-residue modulo every prime in the arithmetic progression $m + 4a$, $m + 8a$, ..., if there *are* any primes in this series, but that wasn't proved until decades after Legendre.

Eisenstein's formula requires relative primality conditions (which are true if the numbers are prime).

Let a,b,a',b' be positive odd integers such that:

$$\gcd(a,b)=\gcd(a',b')=1$$
$$a\equiv a'\pmod 4)$$
$$b\equiv b'\pmod 4)$$

Then

$$\left(\frac{a}{b}\right)\left(\frac{b}{a}\right)=\left(\frac{a'}{b'}\right)\left(\frac{b'}{a'}\right).$$

Hilbert Symbol

The quadratic reciprocity law can be formulated in terms of the Hilbert symbol $(a,b)_v$ where a and b are any two nonzero rational numbers and v runs over all the non-trivial absolute values of the rationals (the archimedean one and the p-adic absolute values for primes p). The Hilbert symbol $(a,b)_v$ is 1 or −1. It is defined to be 1 if and only if the equation $ax^2+by^2=z^2$ has a solution in the completion of the rationals at v other than $x=y=z=0$. The Hilbert reciprocity law states that (a,b) , for fixed a and b and varying v, is 1 for all but finitely many v and the product of $(a,b)_v$ over all v is 1. (This formally resembles the residue theorem from complex analysis).

The proof of Hilbert reciprocity reduces to checking a few special cases, and the non-trivial cases turn out to be equivalent to the main law and the two supplementary laws of quadratic reciprocity for the Legendre symbol. There is no kind of reciprocity in the Hilbert reciprocity law; its name simply indicates the historical source of the result in quadratic reciprocity. Unlike quadratic reciprocity, which requires sign conditions (namely positivity of the primes involved) and a special treatment of the prime 2, the Hilbert reciprocity law treats all absolute values of the rationals on an equal footing. Therefore, it is a more natural way of expressing quadratic reciprocity with a view towards generalization: the Hilbert reciprocity law extends with very few changes to all global fields and this extension can rightly be considered a generalization of quadratic reciprocity to all global fields.

Connection with Cyclotomic Fields

The early proofs of quadratic reciprocity are relatively unilluminating. The situation changed when Gauss used Gauss sums to show that quadratic fields are subfields of cyclotomic fields, and implicitly deduced quadratic reciprocity from a reciprocity theorem for cyclotomic fields. His proof was cast in modern form by later algebraic number theorists. This proof served as a template for class field theory, which can be viewed as a vast generalization of quadratic reciprocity.

Robert Langlands formulated the Langlands program, which gives a conjectural vast generalization of class field theory.

Other Rings

There are also quadratic reciprocity laws in rings other than the integers.

Gaussian Integers

In his second monograph on quartic reciprocity Gauss stated quadratic reciprocity for the ring $\mathbb{Z}[i]$ of Gaussian integers, saying that it is a corollary of the biquadratic law in $\mathbb{Z}[i]$ but did not provide a proof of either theorem. Dirichlet showed that the law in $\mathbb{Z}[i]$ can be deduced from the law for $\mathbb{Z}$ without using biquadratic reciprocity.

For an odd Gaussian prime π and a Gaussian integer α relatively prime to π define the quadratic character for $\mathbb{Z}[i]$ by:

$$\left[\frac{\alpha}{\pi}\right]_2 \equiv \alpha^{\frac{N\pi-1}{2}} \pmod{\pi} = \begin{cases} 1 & \exists \eta \in \mathbb{Z}[i] : \alpha \equiv \eta^2 \pmod{\pi} \\ -1 & \text{otherwise} \end{cases}.$$

Let $\lambda = a + bi, \mu = c + di$ be distinct Gaussian primes where a and c are odd and b and d are even. Then

$$\left[\frac{\lambda}{\mu}\right]_2 = \left[\frac{\mu}{\lambda}\right]_2, \quad \left[\frac{i}{\lambda}\right]_2 = (-1)^{\frac{b}{2}}, \quad \left[\frac{1+i}{\lambda}\right]_2 = \left(\frac{2}{a+b}\right).$$

Eisenstein Integers

Consider the following third root of unity:

$$\omega = \frac{-1+\sqrt{-3}}{2} = e^{\frac{2\pi \acute{y}}{3}}.$$

The ring of Eisenstein integers is $\mathbb{Z}[\omega]$. For an Eisenstein prime $\pi, N\pi \neq 3$, and an Eisenstein integer α with $\gcd(\alpha, \pi) = 1$, define the quadratic character for $\mathbb{Z}[\omega]$ by the formula

$$\left[\frac{\alpha}{\pi}\right]_2 \equiv \alpha^{\frac{N\pi-1}{2}} \pmod{\pi} = \begin{cases} 1 & \exists \eta \in \mathbb{Z}[\omega] : \alpha \equiv \eta^2 \pmod{\pi} \\ -1 & \text{otherwise} \end{cases}.$$

Let $\lambda = a + b\omega$ and $\mu = c + d\omega$ be distinct Eisenstein primes where a and c are not divisible by 3 and b and d are divisible by 3. Eisenstein proved

$$\left[\frac{\lambda}{\mu}\right]_2 \left[\frac{\mu}{\lambda}\right]_2 = (-1)^{\frac{N\lambda-1}{2}\frac{N\mu-1}{2}}, \quad \left[\frac{1-\omega}{\lambda}\right]_2 = \left(\frac{a}{3}\right), \quad \left[\frac{2}{\lambda}\right]_2 = \left(\frac{2}{N\lambda}\right).$$

Imaginary Quadratic Fields

The above laws are special cases of more general laws that hold for the ring of integers in any imaginary quadratic number field. Let k be an imaginary quadratic number field with ring of integers $\mathcal{O}_k$. For a prime ideal $\mathfrak{p} \subset \mathcal{O}_k$ with odd norm $N\mathfrak{p}$ and $\alpha \in \mathcal{O}_k$, define the quadratic character for $\mathcal{O}_k$ as

$$\left[\frac{\alpha}{\mathfrak{p}}\right]_2 \equiv \alpha^{\frac{N\mathfrak{p}-1}{2}} \pmod{\mathfrak{p}} = \begin{cases} 1 & \alpha \notin \mathfrak{p} \text{ and } \exists \eta \in \mathcal{O}_k \text{ such that } \alpha - \eta^2 \in \mathfrak{p} \\ -1 & \alpha \notin \mathfrak{p} \text{ and there is no such } \eta \\ 0 & \alpha \in \mathfrak{p} \end{cases}$$

for an arbitrary ideal $\mathfrak{a} \subset \mathcal{O}_k$ factored into prime ideals $\mathfrak{a} = \mathfrak{p}_1 \cdots \mathfrak{p}_n$ define

$$\left[\frac{\alpha}{\mathfrak{a}}\right]_2 = \left[\frac{\alpha}{\mathfrak{p}_1}\right]_2 \cdots \left[\frac{\alpha}{\mathfrak{p}_n}\right]_2,$$

and for $\beta \in \mathcal{O}_k$ define

$$\left[\frac{\alpha}{\beta}\right]_2 = \left[\frac{\alpha}{\beta\mathcal{O}_k}\right]_2.$$

Let $\mathcal{O}_k = \mathbb{Z}\omega_1 \oplus \mathbb{Z}\omega_2$, i.e. $\{\omega_1, \omega_2\}$ is an integral basis for $\mathcal{O}_k$. For $\nu \in \mathcal{O}_k$ with odd norm $N\nu$, define (ordinary) integers a, b, c, d by the equations,

$$\nu\omega_1 = a\omega_1 + b\omega_2$$
$$\nu\omega_2 = c\omega_1 + d\omega_2$$

and a function

$$\chi(\nu) := i^{(b^2-a+2)c+(a^2-b+2)d+ad}.$$

If $m = N\mu$ and $n = N\nu$ are both odd, Herglotz proved

$$\left[\frac{\mu}{\nu}\right]_2 \left[\frac{\nu}{\mu}\right]_2 = (-1)^{\frac{m-1}{2}\frac{n-1}{2}} \chi(\mu)^{m\frac{n-1}{2}} \chi(\nu)^{-n\frac{m-1}{2}}.$$

Also, if

$$\mu \equiv \mu' \pmod 4 \quad \textit{and} \quad \nu \equiv \nu' \pmod 4$$

Then

$$\left[\frac{\mu}{\nu}\right]_2 \left[\frac{\nu}{\mu}\right]_2 = \left[\frac{\mu'}{\nu'}\right]_2 \left[\frac{\nu'}{\mu'}\right]_2.$$

Polynomials over a Finite Field

Let F be a finite field with $q = p^n$ elements, where p is an odd prime number and n is positive, and let $F[x]$ be the ring of polynomials in one variable with coefficients in F. If $f, g \in F[x]$ and f is irreducible, monic, and has positive degree, define the quadratic character for $F[x]$ in the usual manner:

$$\left(\frac{g}{f}\right) = \begin{cases} 1 & \gcd(f,g) = 1 \text{ and } \exists h,k \in F[x] \text{ such that } g - h^2 = kf \\ -1 & \gcd(f,g) = 1 \text{ and } g \text{ is not a square} \pmod f \\ 0 & \gcd(f,g) \neq 1 \end{cases}$$

If $f = f_1 \cdots f_n$ is a product of monic irreducibles let

$$\left(\frac{g}{f}\right) = \left(\frac{g}{f_1}\right) \cdots \left(\frac{g}{f_n}\right).$$

Dedekind proved that if $f, g \in F[x]$ are monic and have positive degrees,

$$\left(\frac{g}{f}\right)\left(\frac{f}{g}\right) = (-1)^{\frac{q-1}{2}(\deg f)(\deg g)}.$$

ROOT OF UNITY

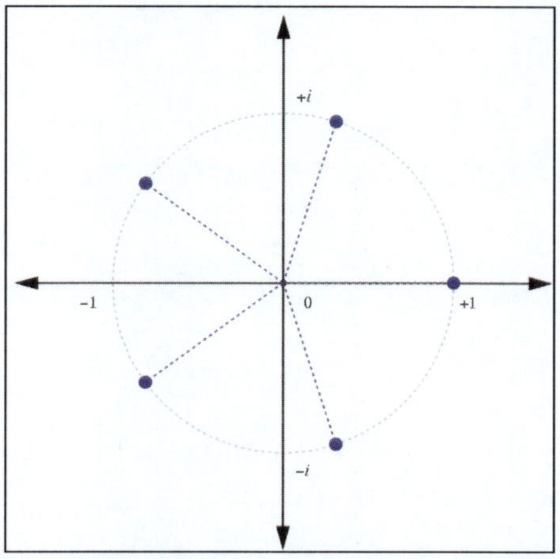

The 5th roots of unity (blue points) in the complex plane.

In mathematics, a root of unity, occasionally called a de Moivre number, is any complex number that yields 1 when raised to some positive integer power n. Roots of unity are used in many branches of mathematics, and are especially important in number theory, the theory of group characters, and the discrete Fourier transform.

Roots of unity can be defined in any field. If the characteristic of the field is zero, they are complex numbers that are also algebraic integers. In positive characteristic, they belong to a finite field, and, conversely, every nonzero element of a finite field is a root of unity. Any algebraically closed field contains exactly n nth roots of unity, except if n is a multiple of the (positive) characteristic of the field.

An *nth root of unity*, where n is a positive integer (i.e. $n = 1, 2, 3, \ldots$), is a number z satisfying the equation

$$z^n = 1.$$

Unless otherwise specified, the roots of unity may be taken to be complex numbers (including the number 1, and the number −1 if n is even, which are complex with a zero imaginary part), and in this case, the nth roots of unity are

$$\exp\left(\frac{2k\pi i}{n}\right) = \cos\frac{2k\pi}{n} + i\sin\frac{2k\pi}{n}, \qquad k = 0, 1, \ldots, n-1.$$

However, the defining equation of roots of unity is meaningful over any field (and even over any ring) F, and this allows considering roots of unity in F. Whichever is the field F, the roots of unity in F are either complex numbers, if the characteristic of F is 0, or, otherwise, belong to a finite field. Conversely, every nonzero element in a finite field is a root of unity in that field.

An nth root of unity is said to be primitive if it is not a kth root of unity for some smaller k, that is if

$$z^n = 1 \quad and \quad z^k \neq 1 \, for \; k = 1,2,3,\ldots,n-1.$$

If n is a prime number, all nth roots of unity, except 1, are primitive.

In the above formula in terms of exponential and trigonometric functions, the primitive nth roots of unity are those for which k and n are coprime integers.

Elementary Properties

Every nth root of unity z is a primitive ath root of unity for some $a \le n$, which is the smallest positive integer such that $z^a = 1$.

Any integer power of an nth root of unity is also an nth root of unity, as

$$(z^k)^n = z^{kn} = (z^n)^k = 1^k = 1.$$

This is also true for negative exponents. In particular, the reciprocal of an nth root of unity is its complex conjugate, and is also an nth root of unity:

$$\frac{1}{z} = z^{-1} = z^{n-1} = \overline{z}.$$

If z is an nth root of unity and $a \equiv b \pmod{n}$ then $z^a = z^b$. In fact, by the definition of congruence, $a = b + kn$ for some integer k, and

$$z^a = z^{b+kn} = z^b z^{kn} = z^b (z^n)^k = z^b 1^k = z^b.$$

Therefore, given a power z^a of z, one has $z^a = z^r$, where $0 \le r < n$ is the remainder of the Euclidean division of a by n.

Let z be a primitive nth root of unity. Then the powers $z, z^2, \ldots, z^{n-1}, z^n = z^0 = 1$ are nth root of unity and are all distinct. (If $z^a = z^b$ where $1 \le a < b \le n$, then $z^{b-a} = 1$, which would imply that z would not be primitive). This implies that $z, z^2, \ldots, z^{n-1}, z^n = z^0 = 1$ are all of the nth roots of unity, since an nth-degree polynomial equation has at most n distinct solutions.

From the preceding, it follows that, if z is a primitive nth root of unity, then $z^a = z^b$ if and only if $a \equiv b(\bmod n)$. If z is not primitive then $a \equiv b(\bmod n)$ implies $z^a = z^b$, but the converse may be false, as shown by the following example. If $n = 4$, a non-primitive nth root of unity is $z = -1$, and one has $z^2 = z^4 = 1$, although $2 \not\equiv 4(\bmod 4)$.

Let z be a primitive nth root of unity. A power $w = z^k$ of z is a primitive ath root of unity for

$$a = \frac{n}{\gcd(k,n)},$$

where $\gcd(k,n)$ is the greatest common divisor of n and k. This results from the fact that ka is the smallest multiple of k that is also a multiple of n. In other words, ka is the least common multiple of k and n. Thus,

$$a = \frac{\operatorname{lcm}(k,n)}{k} = \frac{kn}{k\gcd(k,n)} = \frac{n}{\gcd(k,n)}.$$

Thus, if k and n are coprime, z^k is also a primitive nth root of unity, and therefore there are $\varphi(n)$ (where φ is Euler's totient function) distinct primitive nth roots of unity. (This implies that if n is a prime number, all the roots except +1 are primitive.)

In other words, if $\mathrm{R}(n)$ is the set of all nth roots of unity and $\mathrm{P}(n)$ is the set of primitive ones, $\mathrm{R}(n)$ is a disjoint union of the $\mathrm{P}(n)$:

$$\mathrm{R}(n) = \bigcup_{d|n} \mathrm{P}(d),$$

where the notation means that d goes through all the divisors of n, including 1 and n.

Since the cardinality of $\mathrm{R}(n)$ is n, and that of $\mathrm{P}(n)$ is $\varphi(n)$, this demonstrates the classical formula,

$$\sum_{d|n} \varphi(d) = n.$$

Group Properties

Group of all Roots of Unity

The product and the multiplicative inverse of two roots of unity are also roots of unity. In fact, if $x^m = 1$ and $y^n = 1$, then $(x^{-1})^m = 1$, and $(xy)^k = 1$, where k is the least common multiple of m and n.

Therefore, the roots of unity form an abelian group under multiplication. This group is the torsion subgroup of the circle group.

Group of *n*th Roots of Unity

The product and the multiplicative inverse of two nth roots of unity are also nth roots of unity. Therefore, the nth roots of unity form a group under multiplication.

Given a primitive nth root of unity ω, the other nth roots are powers of ω. This means that the group of the nth roots of unity is a cyclic group. It is worth remarking that the term of *cyclic group* originated from the fact that this group is a subgroup of the circle group.

Galois Group of the Primitive *n*th Roots of Unity

Let $\mathbb{Q}(\omega)$ be the field extension of the rational numbers generated over $\mathbb{Q}$ by a primitive nth root of unity ω. As every nth root of unity is a power of ω, the field $\mathbb{Q}(\omega)$ contains all nth roots of unity, and $\mathbb{Q}(\omega)$ is a Galois extension of $\mathbb{Q}$.

If k is an integer, ω^k is a primitive nth root of unity if and only if k and n are coprime. In this case, the map $\omega \mapsto \omega^k$ induces an automorphism of $\mathbb{Q}(\omega)$, which maps every nth root of unity to its kth power. Every automorphism of $\mathbb{Q}(\omega)$ is obtained in this way, and these automorphisms form the Galois group of $\mathbb{Q}(\omega)$ over the field of the rationals.

The rules of exponentiation imply that the composition of two such automorphisms is obtained by multiplying the exponents. It follows that the map $k \mapsto \left(\omega \mapsto \omega^k\right)$ defines a group isomorphism between the units of the ring of integers modulo n and the Galois group of $\mathbb{Q}(\omega)$.

This shows that this Galois group is abelian, and implies thus that the primitive roots of unity may be expressed in terms of radicals.

Trigonometric Expression

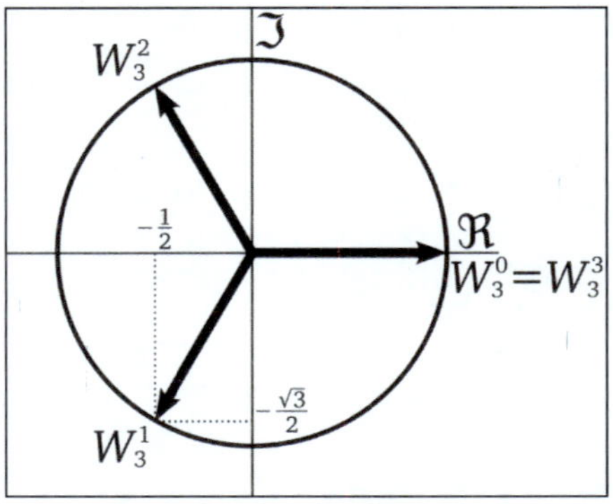

The 3rd roots of unity.

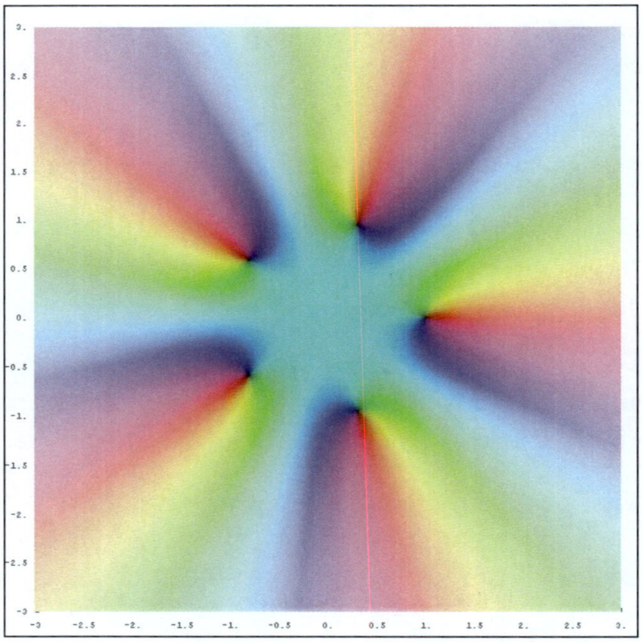

Plot of $z^5 - 1$, in which a zero is represented by the color black.

De Moivre's formula, which is valid for all real x and integers n, is

$$\left(\cos x + i \sin x\right)^n = \cos nx + i \sin nx.$$

Setting $x = \frac{2\pi}{n}$ gives a primitive nth root of unity, one gets

$$\left(\cos\frac{2\pi}{n} + i\sin\frac{2\pi}{n}\right)^n = \cos 2\pi + i\sin 2\pi = 1,$$

but

$$\left(\cos\frac{2\pi}{n} + i\sin\frac{2\pi}{n}\right)^k = \cos\frac{2k\pi}{n} + i\sin\frac{2k\pi}{n} \neq 1$$

for $k = 1, 2, \ldots, n - 1$. In other words,

$$\cos\frac{2\pi}{n} + i\sin\frac{2\pi}{n}$$

is a primitive nth root of unity.

This formula shows that on the complex plane the nth roots of unity are at the vertices of a regular n-sided polygon inscribed in the unit circle, with one vertex at 1. This geometric fact accounts for the term "cyclotomic" in such phrases as cyclotomic field and cyclotomic polynomial; it is from the Greek roots "cyclo" (circle) plus "tomos" (cut, divide).

Euler's formula:

$$e^{ix} = \cos x + i\sin x,$$

which is valid for all real x, can be used to put the formula for the nth roots of unity into the form

$$e^{2\pi i\frac{k}{n}} \qquad 0 \leq k < n.$$

This is a primitive nth-root if and only if the fraction $\frac{k}{n}$ is in lowest terms, i.e. that k and n are coprime.

Algebraic Expression

The nth roots of unity are, by definition, the roots of the polynomial $x^n - 1$, and are thus algebraic numbers. As this polynomial is not irreducible (except for $n = 1$), the primitive nth roots of unity are roots of an irreducible polynomial of lower degree, called the cyclotomic polynomial, and often denoted Φ_n. The degree of Φ_n is given by Euler's totient function, which counts (among other things) the number of primitive nth roots of unity. The roots of Φ_n are exactly the primitive nth roots of unity.

Galois theory can be used to show that cyclotomic polynomials may be conveniently solved in terms of radicals. (The trivial form $\sqrt[n]{1}$ is not convenient, because it contains non-primitive roots, such as 1, which are not roots of the cyclotomic polynomial, and because it does not give the real and imaginary parts separately.) This means that, for each positive integer n, there exists an expression built from integers by root extractions, additions, subtractions, multiplication, and divisions (nothing else), such that the primitive nth roots of unity are exactly the set of values that can be obtained by choosing values for the root extractions (k possible values for a kth root).

Gauss proved that a primitive nth root of unity can be expressed using only square roots, addition, subtraction, multiplication and division if and only if it is possible to construct with compass and straightedge the regular n-gon. This is the case if and only if n is either a power of two or the product of a power of two and Fermat primes that are all different.

If z is a primitive nth root of unity, the same is true for $1/z$, and $r = z + \frac{1}{z}$ is twice the real part of z. In other words, Φ_n is a reciprocal polynomial, the polynomial R_n that has r as a root may be deduced from Φ_n by the standard manipulation on reciprocal polynomials, and the primitive nth roots of unity may be deduced from the roots of R_n by solving the quadratic equation $z^2 - rz + 1 = 0$. That is, the real part of the primitive root is $\frac{r}{2}$, and its imaginary part is $\pm i\sqrt{1-\left(\frac{r}{2}\right)^2}$.

The polynomial R_n is an irreducible polynomial whose all roots are real. Its degree is a power of two, if and only if n is a product of a power of two by a product (possibly empty) of distinct Fermat primes, and the regular n-gon is constructible with compass and straightedge. Otherwise, it is solvable in radicals, but one are in the casus irreducibilis, that is, every expression of the roots in terms of radicals involves *nonreal radicals*.

Explicit Expressions in Low Degrees

- For $n = 1$, the cyclotomic polynomial is $\Phi_1(x) = x - 1$. Therefore, the only primitive first root of unity is 1, which is a non-primitive nth root of unity for every n greater than 1.
- As $\Phi_2(x) = x + 1$, the only primitive second (square) root of unity is -1, which is also a non-primitive nth root of unity for every even $n > 2$. With the preceding case, this completes the list of real roots of unity.
- As $\Phi_3(x) = x^2 + x + 1$, the primitive third (cube) roots of unity, which are the roots of this quadratic polynomial, are

$$\frac{-1+i\sqrt{3}}{2}, \frac{-1-i\sqrt{3}}{2}.$$

- As $\Phi_4(x) = x^2 + 1$, the two primitive fourth roots of unity are i and $-i$.
- As $\Phi_5(x) = x^4 + x^3 + x^2 + x + 1$, the four primitive fifth roots of unity are the roots of this quartic polynomial, which may be explicitly solved in terms of radicals, giving the roots

$$\frac{\varepsilon\sqrt{5}-1}{4} \pm i\frac{\sqrt{10+2\varepsilon\sqrt{5}}}{4},$$

where ε may take the two values 1 and -1 (the same value in the two occurrences).

- As $\Phi_6(x) = x^2 - x + 1$, there are two primitive sixth roots of unity, which are the negatives (and also the square roots) of the two primitive cube roots:

  $$\frac{1+i\sqrt{3}}{2}, \frac{1-i\sqrt{3}}{2}.$$

- As 7 is not a Fermat prime, the seventh roots of unity are the first that require cube roots. There are 6 primitive seventh roots of unity, which are pairwise complex conjugate. The sum of a root and its conjugate is twice its real part. These three sums are the three real roots of the cubic polynomial $r^3 + r^2 - 2r - 1$, and the primitive seventh roots of unity are

  $$\frac{r}{2} \pm i\sqrt{1 - \frac{r^2}{4}},$$

  where r runs over the roots of the above polynomial. As for every cubic polynomial, these roots may be expressed in terms of square and cube roots. However, as these three roots are all real, this is casus irreducibilis, and any such expression involves nonreal cube roots.

- As $\Phi_8(x) = x^4 + 1$, the four primitive eighth roots of unity are the square roots of the primitive fourth roots, $\pm i$. They are thus,

  $$\pm\frac{\sqrt{2}}{2} \pm i\frac{\sqrt{2}}{2}.$$

Periodicity

If z is a primitive nth root of unity, then the sequence of powers

$$\ldots, z^{-1}, z^0, z^1, \ldots$$

is n-periodic (because $z^{j+n} = z^j \cdot z^n = z^j \cdot 1 = z^j$ for all values of j), and the n sequences of powers

$$s_k: \ldots, z^{k\cdot(-1)}, z^{k\cdot 0}, z^{k\cdot 1}, \ldots$$

for $k = 1, \ldots, n$ are all n-periodic (because $z^{k\cdot(j+n)} = z^{k\cdot j}$). Furthermore, the set $\{s_1, \ldots, s_n\}$ of these sequences is a basis of the linear space of all n-periodic sequences. This means that *any* n-periodic sequence of complex numbers

$$\ldots, x_{-1}, x_0, x_1, \ldots$$

can be expressed as a linear combination of powers of a primitive nth root of unity:

$$x_j = \sum_k X_k \cdot z^{k\cdot j} = X_1 z^{1\cdot j} + \cdots + X_n \cdot z^{n\cdot j}$$

for some complex numbers $X_1, \ldots, X_n$ and every integer j.

This is a form of Fourier analysis. If j is a (discrete) time variable, then k is a frequency and X_k is a complex amplitude.

Choosing for the primitive nth root of unity

$$z = e^{\frac{2\pi i}{n}} = \cos\frac{2\pi}{n} + i\sin\frac{2\pi}{n}$$

allows x_j to be expressed as a linear combination of cos and sin:

$$x_j = \sum_k A_k \cos\frac{2\pi jk}{n} + \sum_k B_k \sin\frac{2\pi jk}{n}.$$

This is a discrete Fourier transform.

Summation

Let SR(n) be the sum of all the nth roots of unity, primitive or not. Then

$$\mathrm{SR}(n) = \begin{cases} 1, & n = 1 \\ 0, & n > 1. \end{cases}$$

This is an immediate consequence of Vieta's formulas. In fact, the nth roots of unity being the roots of the polynomial X^n-1, their sum is the coefficient of degree $n - 1$, which is either 1 or 0 according whether $n = 1$ or $n > 1$.

Alternatively, for $n = 1$ there is nothing to prove. For $n > 1$ there exists a root $z \neq 1$. Since the set S of all the nth roots of unity is a group, $zS = S$, so the sum satisfies z SR(n) = SR(n), whence SR(n) = 0.

Let SP(n) be the sum of all the primitive nth roots of unity. Then

$$\mathrm{SP}(n) = \mu(n),$$

where $\mu(n)$ is the Möbius function.

In the topic Elementary properties, it was shown that if R(n) is the set of all nth roots of unity and P(n) is the set of primitive ones, R(n) is a disjoint union of the P(n):

$$\mathrm{R}(n) = \bigcup_{d|n} \mathrm{P}(d),$$

This implies

$$\mathrm{SR}(n) = \sum_{d|n} \mathrm{SP}(d).$$

Applying the Möbius inversion formula gives

$$\mathrm{SP}(n) = \sum_{d|n} \mu(d)\mathrm{SR}\left(\frac{n}{d}\right).$$

In this formula, if $d < n$, then $\mathrm{SR}\left(\frac{n}{d}\right) = 0$, and for $d = n$: $\mathrm{SR}\left(\frac{n}{d}\right) = 1$. Therefore, SP($n$) = $\mu(n)$.

This is the special case $c_n(1)$ of Ramanujan's sum $c_n(s)$, defined as the sum of the sth powers of the primitive nth roots of unity:

$$c_n(s) = \sum_{\substack{a=1 \\ \gcd(a,n)=1}}^{n} e^{2\pi i \frac{a}{n} s}.$$

Orthogonality

From the summation formula follows an orthogonality relationship: for $j = 1, \dots, n$ and $j' = 1, \dots, n$

$$\sum_{k=1}^{n} \overline{z^{j\cdot k}} \cdot z^{j'\cdot k} = n \cdot \delta_{j,j'},$$

where δ is the Kronecker delta and z is any primitive nth root of unity.

The $n \times n$ matrix U whose (j, k)th entry is,

$$U_{j,k} = n^{-\frac{1}{2}} \cdot z^{j\cdot k},$$

defines a discrete Fourier transform. Computing the inverse transformation using gaussian elimination requires $O(n^3)$ operations. However, it follows from the orthogonality that U is unitary. That is,

$$\sum_{k=1}^{n} \overline{U_{j,k}} \cdot U_{k,j'} = \delta_{j,j'},$$

and thus the inverse of U is simply the complex conjugate. (This fact was first noted by Gauss when solving the problem of trigonometric interpolation). The straightforward application of U or its inverse to a given vector requires $O(n^2)$ operations. The fast Fourier transform algorithms reduces the number of operations further to $O(n \log n)$.

Cyclotomic Polynomials

The zeroes of the polynomial $p(z) = z^n - 1$ are precisely the nth roots of unity, each with multiplicity 1. The nth *cyclotomic polynomial* is defined by the fact that its zeros are precisely the *primitive* nth roots of unity, each with multiplicity 1.

$$\Phi_n(z) = \prod_{k=1}^{\varphi(n)} (z - z_k)$$

where $z_1, z_2, z_3, \dots, z_{\varphi(n)}$ are the primitive nth roots of unity, and $\varphi(n)$ is Euler's totient function. The polynomial $\Phi_n(z)$ has integer coefficients and is an irreducible polynomial over the rational numbers (i.e., it cannot be written as the product of two positive-degree polynomials with rational coefficients). The case of prime n, which is easier than the general assertion, follows by applying Eisenstein's criterion to the polynomial

$$\frac{(z+1)^n-1}{(z+1)-1},$$

and expanding via the binomial theorem.

Every nth root of unity is a primitive dth root of unity for exactly one positive divisor d of n. This implies that,

$$z^n-1=\prod_{d\mid n}\Phi_d(z).$$

This formula represents the factorization of the polynomial $z^n - 1$ into irreducible factors.

$z^1 - 1 = z - 1$

$z^2 - 1 = (z - 1)\cdot(z + 1)$

$z^3 - 1 = (z - 1)\cdot(z^2 + z + 1)$

$z^4 - 1 = (z - 1)\cdot(z + 1)\cdot(z^2 + 1)$

$z^5 - 1 = (z - 1)\cdot(z^4 + z^3 + z^2 + z + 1)$

$z^6 - 1 = (z - 1)\cdot(z + 1)\cdot(z^2 + z + 1)\cdot(z^2 - z + 1)$

$z^7 - 1 = (z - 1)\cdot(z^6 + z^5 + z^4 + z^3 + z^2 + z + 1)$

$z^8 - 1 = (z - 1)\cdot(z + 1)\cdot(z^2 + 1)\cdot(z^4 + 1)$

Applying Möbius inversion to the formula gives

$$\Phi_n(z)=\prod_{d\mid n}\left(z^{\frac{n}{d}}-1\right)^{\mu(d)}=\prod_{d\mid n}\left(z^d-1\right)^{\mu\left(\frac{n}{d}\right)},$$

where μ is the Möbius function.

So the first few cyclotomic polynomials are

$\Phi_1(z) = z - 1$

$\Phi_2(z) = (z^2 - 1)\cdot(z - 1)^{-1} = z + 1$

$\Phi_3(z) = (z^3 - 1)\cdot(z - 1)^{-1} = z^2 + z + 1$

$\Phi_4(z) = (z^4 - 1)\cdot(z^2 - 1)^{-1} = z^2 + 1$

$\Phi_5(z) = (z^5 - 1)\cdot(z - 1)^{-1} = z^4 + z^3 + z^2 + z + 1$

$\Phi_6(z) = (z^6 - 1)\cdot(z^3 - 1)^{-1}\cdot(z^2 - 1)^{-1}\cdot(z - 1) = z^2 - z + 1$

$\Phi_7(z) = (z^7 - 1)\cdot(z - 1)^{-1} = z^6 + z^5 + z^4 + z^3 + z^2 + z + 1$

$\Phi_8(z) = (z^8 - 1)\cdot(z^4 - 1)^{-1} = z^4 + 1$

If p is a prime number, then all the pth roots of unity except 1 are primitive pth roots, and we have

$$\Phi_p(z) = \frac{z^p - 1}{z - 1} = \sum_{k=0}^{p-1} z^k.$$

Substituting any positive integer ≥ 2 for z, this sum becomes a base z repunit. Thus a necessary (but not sufficient) condition for a repunit to be prime is that its length be prime.

Contrary to first appearances, *not* all coefficients of all cyclotomic polynomials are 0, 1, or −1. The first exception is Φ_{105}. It is not a surprise it takes this long to get an example, because the behavior of the coefficients depends not so much on n as on how many odd prime factors appear in n. More precisely, it can be shown that if n has 1 or 2 odd prime factors (e.g., $n = 150$) then the nth cyclotomic polynomial only has coefficients 0, 1 or −1. Thus the first conceivable n for which there could be a coefficient besides 0, 1, or −1 is a product of the three smallest odd primes, and that is $3 \cdot 5 \cdot 7 = 105$. This by itself doesn't prove the 105th polynomial has another coefficient, but does show it is the first one which even has a chance of working (and then a computation of the coefficients shows it does). A theorem of Schur says that there are cyclotomic polynomials with coefficients arbitrarily large in absolute value. In particular, if $n = p_1 \cdot p_2 \cdot \cdots \cdot p_t$, where $p_1 < p_2 < \cdots < p_t$ are odd primes, $p_1 + p_2 > p_t$, and t is odd, then $1 - t$ occurs as a coefficient in the nth cyclotomic polynomial.

Many restrictions are known about the values that cyclotomic polynomials can assume at integer values. For example, if p is prime, then $d \mid \Phi_p(d)$ if and only $d \equiv 1 \pmod{p}$.

Cyclotomic polynomials are solvable in radicals, as roots of unity are themselves radicals. Moreover, there exist more informative radical expressions for nth roots of unity with the additional property that every value of the expression obtained by choosing values of the radicals (for example, signs of square roots) is a primitive nth root of unity. This was already shown by Gauss in 1797. Efficient algorithms exist for calculating such expressions.

Cyclic Groups

The nth roots of unity form under multiplication a cyclic group of order n, and in fact these groups comprise all of the finite subgroups of the multiplicative group of the complex number field. A generator for this cyclic group is a primitive nth root of unity.

The nth roots of unity form an irreducible representation of any cyclic group of order n. The orthogonality relationship also follows from group-theoretic principles as described in character group.

The roots of unity appear as entries of the eigenvectors of any circulant matrix, i.e. matrices that are invariant under cyclic shifts, a fact that also follows from group representation theory as a variant of Bloch's theorem. In particular, if a circulant Hermitian matrix is considered (for example, a discretized one-dimensional Laplacian with periodic boundaries), the orthogonality property immediately follows from the usual orthogonality of eigenvectors of Hermitian matrices.

Cyclotomic Fields

By adjoining a primitive nth root of unity to Q, one obtains the nth cyclotomic field $\mathrm{Q}(\exp(2\pi i/n))$.

This field contains all nth roots of unity and is the splitting field of the nth cyclotomic polynomial over Q. The field extension $Q(\exp(2\pi i/n))/Q$ has degree $\varphi(n)$ and its Galois group is naturally isomorphic to the multiplicative group of units of the ring Z/nZ.

As the Galois group of $Q(\exp(2\pi i/n))/Q$ is abelian, this is an abelian extension. Every subfield of a cyclotomic field is an abelian extension of the rationals. It follows that every nth root of unity may be expressed in term of k-roots, with various k not exceeding $\varphi(n)$. In these cases Galois theory can be written out explicitly in terms of Gaussian periods: this theory from the *Disquisitiones Arithmeticae* of Gauss was published many years before Galois.

Conversely, *every* abelian extension of the rationals is such a subfield of a cyclotomic field – this is the content of a theorem of Kronecker, usually called the *Kronecker–Weber theorem* on the grounds that Weber completed the proof.

Relation to Quadratic Integers

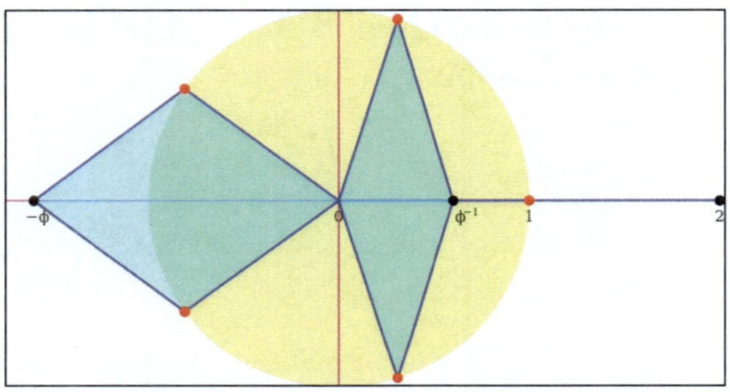

In the complex plane, the red points are the fifth roots of unity, and the blue points are the sums of a fifth root of unity and its complex conjugate.

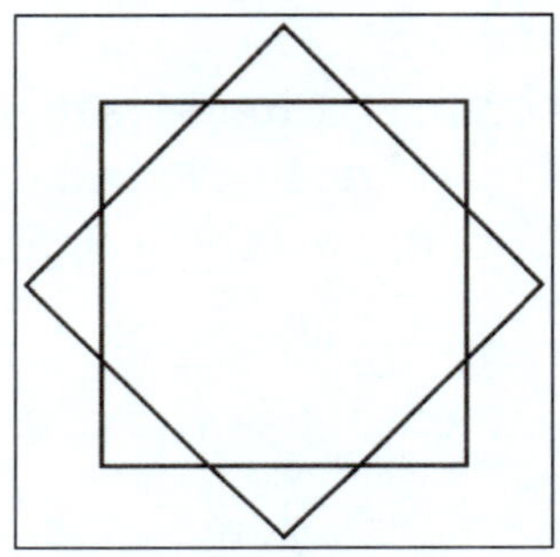

In the complex plane, the corners of the two squares are the eighth roots of unity.

For $n = 1, 2$, both roots of unity 1 and -1 belong to Z.

For three values of n, the roots of unity are quadratic integers:

- For $n = 3, 6$ they are Eisenstein integers ($D = -3$).
- For $n = 4$ they are Gaussian integers ($D = -1$).

For four other values of n, the primitive roots of unity are not quadratic integers, but the sum of any root of unity with its complex conjugate (also an nth root of unity) is a quadratic integer.

For $n = 5, 10$, none of the non-real roots of unity (which satisfy a quartic equation) is a quadratic integer, but the sum $z + z = 2\,\mathrm{Re}\,z$ of each root with its complex conjugate (also a 5th root of unity) is an element of the ring $Z\left[\frac{1+\sqrt{5}}{2}\right]$ ($D = 5$). For two pairs of non-real 5th roots of unity these sums are inverse golden ratio and minus golden ratio.

For $n = 8$, for any root of unity $Z+\overline{Z}$ equals to either 0, ±2, or $\pm\sqrt{2}$ ($D = 2$).

For $n = 12$, for any root of unity, $Z+\overline{Z}$ equals to either 0, ±1, ±2 or $\pm\sqrt{3}$ ($D = 3$).

GAUSSIAN PERIOD

In mathematics, in the area of number theory, a Gaussian period is a certain kind of sum of roots of unity. The periods permit explicit calculations in cyclotomic fields connected with Galois theory and with harmonic analysis (discrete Fourier transform). They are basic in the classical theory called cyclotomy. Closely related is the Gauss sum, a type of exponential sum which is a linear combination of periods.

Given an integer $n > 1$, let H be any subgroup of the multiplicative group

$$G = (\mathbb{Z} / n\mathbb{Z})^{\times}$$

of invertible residues modulo n, and let

$$\zeta = \exp\left(\frac{2\pi i}{n}\right).$$

A Gaussian period P is a sum of the primitive n-th roots of unity ζ^a, where a runs through all of the elements in a fixed coset of H in G.

The definition of P can also be stated in terms of the field trace. We have:

$$P = \mathrm{Tr}_{\mathbb{Q}(\zeta)/L}(\zeta^j),$$

for some subfield L of Q(ζ) and some j coprime to n. This corresponds to the previous definition by identifying G and H with the Galois groups of Q(ζ)/Q and Q(ζ)/L, respectively. The choice of j determines the choice of coset of H in G in the previous definition.

Example:

The situation is simplest when n is a prime number $p > 2$. In that case G is cyclic of order $p - 1$, and has one subgroup H of order d for every factor d of $p - 1$. For example, we can take H of index two. In that case H consists of the quadratic residues modulo p. Corresponding to this H we have the Gaussian period

$$P = \zeta + \zeta^4 + \zeta^9 + \cdots$$

summed over (p – 1)/2 quadratic residues, and the other period P* summed over the (p – 1)/2 quadratic non-residues. It is easy to see that $P + P^* = -1$ since the left-hand side adds all the primitive p-th roots of 1. We also know, from the trace definition, that P lies in a quadratic extension of Q. Therefore, as Gauss knew, P satisfies a quadratic equation with integer coefficients. Evaluating the square of the sum P is connected with the problem of counting how many quadratic residues between 1 and p – 1 are succeeded by quadratic residues. The solution is elementary (as we would now say, it computes a local zeta-function, for a curve that is a conic). One has

$$(P - P^*)^2 = p \text{ or } -p, \text{ for } p = 4m + 1 \text{ or } 4m + 3 \text{ respectively.}$$

This therefore gives us the precise information about which quadratic field lies in **Q**(ζ). (That could be derived also by ramification arguments in algebraic number theory).

As Gauss eventually showed, to evaluate $P - P^*$, the correct square root to take is the positive (resp. i times positive real) one, in the two cases. Thus the explicit value of the period P is given by

$$P = \begin{cases} \dfrac{-1+\sqrt{p}}{2}, & \text{if } p = 4m+1, \\ \dfrac{-1+i\sqrt{p}}{2}, & \text{if } p = 4m+3. \end{cases}$$

Gauss Sums

The Gaussian periods are closely related to another class of sums of roots of unity, now generally called Gauss sums (sometimes Gaussian sums). The quantity $P - P^*$ presented above is a quadratic Gauss sum mod p, the simplest non-trivial example of a Gauss sum. One observes that $P - P^*$ may also be written as:

$$\sum \chi(a)\zeta^a,$$

where $\chi(a)$ here stands for the Legendre symbol (a/p), and the sum is taken over residue classes modulo p. More generally, given a Dirichlet character χ mod n, the Gauss sum mod n associated with χ is:

$$G(k,\chi) = \sum_{m=1}^{n} \chi(m)\exp\left(\frac{2\pi imk}{n}\right).$$

For the special case of $\chi = \chi_1$ the principal Dirichlet character, the Gauss sum reduces to the Ramanujan sum:

$$G(k,\chi_1) = c_n(k) = \sum_{m=1;(m,n)=1}^{n} \exp\left(\frac{2\pi imk}{n}\right) = \sum_{d|(n,k)} d\mu\left(\frac{n}{d}\right)$$

where μ is the Möbius function.

The Gauss sums $G(k,\chi)$ are ubiquitous in number theory; for example they occur significantly in the functional equations of L-functions. Gauss sums are in a sense the finite field analogues of the gamma function.

Relationship of Gaussian Periods and Gauss Sums

The Gaussian periods are related to the Gauss sums $G(1,\chi)$ for which the character χ is trivial on H. Such χ take the same value at all elements a in a fixed coset of H in G. For example, the quadratic character mod p described above takes the value 1 at each quadratic residue, and takes the value -1 at each quadratic non-residue. The Gauss sum $G(1,\chi)$ can thus be written as a linear combination of Gaussian periods (with coefficients $\chi(a)$); the converse is also true, as a consequence of the orthogonality relations for the group $(\mathbb{Z}/n\mathbb{Z})^\times$. In other words, the Gaussian periods and Gauss sums are each other's Fourier transforms. The Gaussian periods generally lie in smaller fields, since for example when n is a prime p, the values $\chi(a)$ are $(p - 1)$-th roots of unity. On the other hand, Gauss sums have nicer algebraic properties.

HASSE NORM THEOREM

In number theory, the Hasse norm theorem states that if L/K is a cyclic extension of number fields, then if a nonzero element of K is a local norm everywhere, then it is a global norm. Here to be a global norm means to be an element *k* of K such that there is an element *l* of L with $N_{L/K}(l) = k$; in other words *k* is a relative norm of some element of the extension field L. To be a local norm means that for some prime p of K and some prime P of L lying over K, then *k* is a norm from L_P; here the "prime" p can be an archimedean valuation, and the theorem is a statement about completions in all valuations, archimedean and non-archimedean.

The theorem is no longer true in general if the extension is abelian but not cyclic. Hasse gave the counterexample that 3 is a local norm everywhere for the extension $Q(\sqrt{-3}, \sqrt{13}) / Q$ but is not a global norm. Serre and Tate showed that another counterexample is given by the field $Q(\sqrt{13}, \sqrt{17}) / Q$ where every rational square is a local norm everywhere but 5^2 is not a global norm.

This is an example of a theorem stating a local-global principle.

The full theorem is due to Hasse. The special case when the degree *n* of the extension is 2 was proved by Hilbert, and the special case when *n* is prime was proved by Furtwangler.

The Hasse norm theorem can be deduced from the theorem that an element of the Galois cohomology group $H^2(L/K)$ is trivial if it is trivial locally everywhere, which is in turn equivalent to the deep theorem that the first cohomology of the idele class group vanishes. This is true for all finite Galois extensions of number fields, not just cyclic ones. For cyclic extensions the group $H^2(L/K)$ is isomorphic to the Tate cohomology group $H^0(L/K)$ which describes which elements are norms, so for cyclic extensions it becomes Hasse's theorem that an element is a norm if it is a local norm everywhere.

CLASS FIELD THEORY

In mathematics, class field theory is the branch of algebraic number theory concerned with the abelian extensions of number fields, global fields of positive characteristic, and local fields. The theory had its origins in the proof of quadratic reciprocity by Gauss at the end of 18th century. These ideas were developed over the next century, giving rise to a set of conjectures by Hilbert that were subsequently proved by Takagi and Artin. These conjectures and their proofs constitute the main body of class field theory.

One major result states that, given a number field *F*, and writing *K* for the maximal abelian unramifed extension of *F*, the Galois group of *K* over *F* is canonically isomorphic to the ideal class group of *F*. This statement can be generalized to the Artin reciprocity law; writing C_F for the idele class group of *F*, and taking *L* to be any finite abelian extension of *F*, this law gives a canonical isomorphism

$$\theta_{L/F} : C_F / N_{L/F}(C_L) \to \text{Gal}(L / F),$$

where $N_{L/F}$ denotes the idelic norm map from L to F. This isomorphism is then called the reciprocity map. The existence theorem states that the reciprocity map can be used to give a bijection between the set of abelian extensions of F and the set of closed subgroups of finite index of C_F.

A standard method for developing global class field theory since the 1930s is to develop local class field theory, which describes abelian extensions of local fields, and then use it to construct global class field theory. This was first done by Artin and Tate using the theory of group cohomology, and in particular by developing the notion of class formations. More recently, Neukrich has found a proof of the main statements of global class field theory without using cohomological ideas.

Class field theory also encompasses the explicit construction of maximal abelian extensions of number fields in the few cases where such constructions are known. Currently, this portion of the theory consists of Kronecker-Weber theorem, which can be used to construct the abelian extensions of $\mathbb{Q}$ and the theory of complex multiplication, which can be used to construct the abelian extensions of CM-fields.

The Langlands program gives one approach for generalizing class field theory to non-abelian extensions. This generalization is mostly still conjectural. For number fields, class field theory and the results related to the modularity theorem are the only cases known.

In modern mathematical language class field theory can be formulated as follows. Consider the maximal abelian extension A of a local or global field K. It is of infinite degree over K; the Galois group G of A over K is an infinite pro-finite group, so a compact topological group, and it is abelian. The central aims of class field theory are: to describe G in terms of certain appropriate topological objects associated to K, to describe finite abelian extensions of K in terms of open subgroups of finite index in the topological object associated to K. In particular, one wishes to establish a one-to-one correspondence between finite abelian extensions of K and their norm groups in this topological object for K. This topological object is the multiplicative group in the case of local fields with finite residue field and the idele class group in the case of global fields. The finite abelian extension corresponding to an open subgroup of finite index is called the class field for that subgroup, which gave the name to the theory.

The fundamental result of general class field theory states that the group G is naturally isomorphic to the profinite completion of C_K, the multiplicative group of a local field or the idele class group of the global field, with respect to the natural topology on C_K related to the specific structure of the field K. Equivalently, for any finite Galois extension L of K, there is an isomorphism (the Artin reciprocity map)

$$\mathrm{Gal}(L/K)^{\mathrm{ab}} \to C_K / N_{L/K}(C_L),$$

of the abelianization of the Galois group of the extension with the quotient of the idele class group of K by the image of the norm of the idele class group of L.

For some small fields, such as the field of rational numbers $\mathbb{Q}$ or its quadratic imaginary extensions there is a more detailed very explicit but too specific theory which provides more information. For example, the abelianized absolute Galois group G of $\mathbb{Q}$ is (naturally isomorphic to) an infinite product of the group of units of the p-adic integers taken over all prime numbers p, and the corresponding maximal abelian extension of the rationals is the field generated by all roots of unity. This is known as the Kronecker–Weber theorem, originally conjectured by Leopold Kronecker. In this case the reciprocity isomorphism of class field theory (or Artin reciprocity map) also admits an explicit description due to the Kronecker–Weber theorem. However, principal constructions of such more detailed theories for small algebraic number fields are not extendable to the general case of algebraic number fields, and different conceptual principles are in use in the general class field theory.

The standard method to construct the reciprocity homomorphism is to first construct the local reciprocity isomorphism from the multiplicative group of the completion of a global field to the Galois group of its maximal abelian extension (this is done inside local class field theory) and then prove that the product of all such local reciprocity maps when defined on the idele group of the global field is trivial on the image of the multiplicative group of the global field. The latter property is called the global reciprocity law and is a far reaching generalization of the Gauss quadratic reciprocity law.

One of the methods to construct the reciprocity homomorphism uses class formation which derives class field theory from axioms of class field theory. This derivation is purely topological group theoretical, while to establish the axioms one has to use the ring structure of the ground field.

There are methods which use cohomology groups, in particular the Brauer group, and there are methods which do not use cohomology groups and are very explicit and fruitful for applications.

Applications

Class field theory is used to prove Artin-Verdier duality. Very explicit class field theory is used in many subareas of algebraic number theory such as Iwasawa theory and Galois modules theory.

Most main achievements in the Langlands correspondence for number fields, the BSD conjecture for number fields, and Iwasawa theory for number fields are using very explicit but narrow class field theory methods or their generalizations. The open question is therefore to use generalizations of general class field theory in these three directions.

Generalizations of Class Field Theory

There are three main generalizations, each of great interest on its own. They are: the Langlands program, anabelian geometry, and higher class field theory.

Often, the Langlands correspondence is viewed as a nonabelian class field theory. If/when fully established, it would contain a certain theory of nonabelian Galois extensions of global fields. However, the Langlands correspondence does not include as much arithmetical information about finite Galois extensions as class field theory does in the abelian case. It also does not include an analog of the existence theorem in class field theory, i.e. the concept of class fields is absent in the Langlands correspondence. There are several other nonabelian theories, local and global, which provide alternative to the Langlands correspondence point of view.

Another generalization of class field theory is anabelian geometry which studies algorithms to restore the original object (e.g. a number field or a hyperbolic curve over it) from the knowledge of its full absolute Galois group of algebraic fundamental group.

Another natural generalization is higher class field theory. It describes abelian extensions of higher local fields and higher global fields. The latter come as function fields of schemes of finite type over integers and their appropriate localization and completions. The theory is referred to as higher local class field theory and higher global class field theory. It uses algebraic K-theory and appropriate Milnor K-groups replace K_1 which is in use in one-dimensional class field theory.

IDEAL CLASS GROUP

In number theory, the ideal class group (or class group) of an algebraic number field K is the quotient group J_K/P_K where J_K is the group of fractional ideals of the ring of integers of K, and P_K is its subgroup of principal ideals. The class group is a measure of the extent to which unique factorization fails in the ring of integers of K. The order of the group, which is finite, is called the class number of K.

The theory extends to Dedekind domains and their field of fractions, for which the multiplicative properties are intimately tied to the structure of the class group. For example, the class group of a Dedekind domain is trivial if and only if the ring is a unique factorization domain.

If R is an integral domain, define a relation ~ on nonzero fractional ideals of R by $I \sim J$ whenever there exist nonzero elements a and b of R such that $(a)I = (b)J$. (Here the notation (a) means the principal ideal of R consisting of all the multiples of a.) It is easily shown that this is an equivalence relation. The equivalence classes are called the *ideal classes* of R. Ideal classes can be multiplied: if $[I]$ denotes the equivalence class of the ideal I, then the multiplication $[I][J] = [IJ]$ is well-defined and commutative. The principal ideals form the ideal class $[R]$ which serves as an identity element for this multiplication. Thus a class $[I]$ has an inverse $[J]$ if and only if there is an ideal J such that IJ is a principal ideal. In general, such a J may not exist and consequently the set of ideal classes of R may only be a monoid.

However, if R is the ring of algebraic integers in an algebraic number field, or more generally a Dedekind domain, the multiplication defined above turns the set of fractional ideal classes into an abelian group, the ideal class group of R. The group property of existence of inverse elements follows easily from the fact that, in a Dedekind domain, every non-zero ideal (except R) is a product of prime ideals.

Properties

The ideal class group is trivial (i.e. has only one element) if and only if all ideals of R are principal. In this sense, the ideal class group measures how far R is from being a principal ideal domain, and hence from satisfying unique prime factorization (Dedekind domains are unique factorization domains if and only if they are principal ideal domains).

The number of ideal classes (the class number of R) may be infinite in general. In fact, every abelian group is isomorphic to the ideal class group of some Dedekind domain. But if R is in fact a ring of algebraic integers, then the class number is always *finite*. This is one of the main results of classical algebraic number theory.

Computation of the class group is hard, in general; it can be done by hand for the ring of integers in an algebraic number field of small discriminant, using Minkowski's bound. This result gives a bound, depending on the ring, such that every ideal class contains an ideal norm less than the bound. In general the bound is not sharp enough to make the calculation practical for fields with large discriminant, but computers are well suited to the task.

The mapping from rings of integers R to their corresponding class groups is functorial, and the class group can be subsumed under the heading of algebraic K-theory, with $K_0(R)$ being the functor assigning to R its ideal class group; more precisely, $K_0(R) = Z \times C(R)$, where $C(R)$ is the class group. Higher K groups can also be employed and interpreted arithmetically in connection to rings of integers.

Relation with the Group of Units

It was remarked above that the ideal class group provides part of the answer to the question of how much ideals in a Dedekind domain behave like elements. The other part of the answer is provided by the multiplicative group of units of the Dedekind domain, since passage from principal ideals to their generators requires the use of units (and this is the rest of the reason for introducing the concept of fractional ideal, as well):

Define a map from $R^{\times}$ to the set of all nonzero fractional ideals of R by sending every element to the principal (fractional) ideal it generates. This is a group homomorphism; its kernel is the group of units of R, and its cokernel is the ideal class group of R. The failure of these groups to be trivial is a measure of the failure of the map to be an isomorphism: that is the failure of ideals to act like ring elements, that is to say, like numbers.

Examples of Ideal Class Groups

- The rings **Z**, **Z**[ω], and **Z**[*i*], where ω is a cube root of 1 and *i* is a fourth root of 1 (i.e. a square root of −1), are all principal ideal domains (and in fact are all Euclidean domains), and so have class number 1: that is, they have trivial ideal class groups.
- If k is a field, then the polynomial ring $k[X_1, X_2, X_3, \ldots]$ is an integral domain. It has a countably infinite set of ideal classes.

Class Numbers of Quadratic Fields

If d is a square-free integer (a product of distinct primes) other than 1, then $Q\left(\sqrt{d}\right)$ is a quadrat ic extension of Q. If $d < 0$, then the class number of the ring R of algebraic integers of $Q\left(\sqrt{d}\right)$ is equal to 1 for precisely the following values of d: $d = -1, -2, -3, -7, -11, -19, -43, -67$, and -163. This result was first conjectured by Gauss and proven by Kurt Heegner, although Heegner's proof was not believed until Harold Stark gave a later proof in 1967. This is a special case of the famous class number problem.

If, on the other hand, $d > 0$, then it is unknown whether there are infinitely many fields $Q\left(\sqrt{d}\right)$ with class number 1. Computational results indicate that there are a great many such fields. However, it is not even known if there are infinitely many number fields with class number 1.

For $d < 0$, the ideal class group of $Q\left(\sqrt{d}\right)$ is isomorphic to the class group of integral binary quadratic forms of discriminant equal to the discriminant of $Q\left(\sqrt{d}\right)$. For $d > 0$, the ideal class group may be half the size since the class group of integral binary quadratic forms is isomorphic to the narrow class group of $Q\left(\sqrt{d}\right)$.

Example of a non-trivial class group:

The quadratic integer ring $R = \mathrm{Z}\left[\sqrt{-5}\right]$ is the ring of integers of $\mathrm{Q}\left(\sqrt{-5}\right)$. It does *not* possess unique factorization; in fact the class group of R is cyclic of order 2. Indeed, the ideal

$$J = \left(2, 1+\sqrt{-5}\right),$$

is not principal, which can be proved by contradiction as follows. R has a norm function $N(a+b\sqrt{-5}) = a^2+5b^2$, which satisfies $N(vv) \quad N(u)N(v)$, and $N(u)=1$ if and only if u is a unit in R. First of all, $J \neq R$, because the quotient ring of R modulo the ideal $(1+\sqrt{-5})$ is isomorphic to $\mathrm{Z}/6\mathrm{Z}$, so that the quotient ring of R modulo J is isomorphic to $\mathrm{Z}/2\mathrm{Z}$. If J were generated by an element x of R, then x would divide both 2 and $1+\sqrt{-5}$. Then the norm $N(x)$ would divide both $N(2)=4$ and $N(1+\sqrt{-5})=6$, so $N(x)$ would divide 2. If $N(x)=1$, then x is a unit, and $J = R$, a contradiction. But $N(x)$ cannot be 2 either, because R has no elements of norm 2, because the Diophantine equation $b^2+5c^2=2$ has no solutions in integers, as it has no solutions modulo 5.

One also computes that $J^2 = (2)$, which is principal, so the class of J in the ideal class group has order two. Showing that there aren't any *other* ideal classes requires more effort.

The fact that this J is not principal is also related to the fact that the element 6 has two distinct factorisations into irreducibles:

$$6 = 2\times 3 = \left(1+\sqrt{-5}\right)\times\left(1-\sqrt{-5}\right).$$

Connections to Class Field Theory

Class field theory is a branch of algebraic number theory which seeks to classify all the abelian extensions of a given algebraic number field, meaning Galois extensions with abelian Galois group. A particularly beautiful example is found in the Hilbert class field of a number field, which can be defined as the maximal unramified abelian extension of such a field. The Hilbert class field L of a number field K is unique and has the following properties:

- Every ideal of the ring of integers of K becomes principal in L, i.e., if I is an integral ideal of K then the image of I is a principal ideal in L.
- L is a Galois extension of K with Galois group isomorphic to the ideal class group of K.

Neither property is particularly easy to prove.

KRONECKER–WEBER THEOREM

The Kronecker-Weber theorem, sometimes known as the Kronecker-Weber-Hilbert theorem, is one of the earliest known results in class field theory.

In layman's terms, the Kronecker-Weber theorem says that cyclotomic extensions of the field Q of rational numbers capture in a very precise way all the abelian extensions of Q. More precisely, the theorem says that every abelian extension of Q is contained in a cyclotomic extension, i.e., that for an arbitrary abelian Galois extension K / Q, there exists an integer n for which $K \subset \mathrm{Q}(\zeta_n)$ where ζ_n is a primitive root of unity.

From a more computational viewpoint, the theorem can be rephrased to say that every algebraic integer whose Galois group is abelian can be expressed as a sum of nth roots of unity for some n.

Incomplete proofs of the theorem were given first by Kronecker and by Weber in 1853 and 1886, respectively. The first complete proof was given by Hilbert in 1896, who later hinged the twelfth of what have come to be known as Hilbert's problems on the desire to find generalizations of the theorem for fields other than Q, whereby sufficient analogues of roots of unity and holomorphic functions in several variables would also be necessary.

References

- Gras, Georges (2005), Class Field Theory: from theory to practice (corrected 2nd printing), Springer Monographs in Mathematics, xiii+507 pages, Berlin, New York: Springer-Verlag, ISBN 978-3-540-44133-5
- Algebraic-number-theory: encyclopediaofmath.org, Retrieved 21 May, 2019
- Narkiewicz, Władysław (2004), Elementary and analytic theory of algebraic numbers, Springer Monographs in Mathematics (3 ed.), Berlin: Springer-Verlag, ISBN 978-3-540-21902-6, MR 2078267
- Roberts, David P. (2001), "Density of cubic field discriminants", Mathematics of Computation, 70 (236): 1699–1705, arxiv:math/9904190, doi:10.1090/s0025-5718-00-01291-6, MR 1836927
- Section 12 of Swinnerton-Dyer, H.P.F. (2001), A brief guide to algebraic number theory, London Mathematical Society Student Texts, 50, Cambridge University Press, ISBN 978-0-521-00423-7, MR 1826558
- Kronecker-webertheorem: wolfram.com, Retrieved 8 January, 2019

Combinatorial Number Theory

CHAPTER 5

Combinatorial number theory is the combination of combinatorics and number theory. It includes restricted sumset, Gilbreath's conjecture, Znám's problem, Szemerédi's theorem, etc. This chapter closely examines about these concepts related to combinatorial number theory to provide an easy understanding of the subject.

Combinatorial number theory can be described as an area of mathematics which is a cross between combinatorics and number theory. More precisely, the area concerns structures of integers (or similar sets), with some number theoretic properties, which can be studied mainly by combinatorial means, rather than, for example, by algebraic methods.

Problems and results in combinatorial number theory really cannot be rigorously classified, but here is an effort at exposition of a few broad categories. These represent just some of the major areas of combinatorial number theory and are by no means intended to represent the field completely.

Divisibility Issues

One basic area of study is related to the relation of divisibility between positive integers and its interplay with addition and ordering. Interpreting a number in terms of its prime factorization often simplifies questions of divisibility: a number is divisible by another if and only if each prime has an exponent in the first number at least as large as its exponent in the second; this formulation facilitates calculation of greatest common divisors and least common multiples. However, this interpretation of divisibility is far from trivializing many problems and results, and a variety of other techniques prove essential as well in the study of the seemingly simple notions of divisibility, relative primality, and so forth. Because the set of positive integers, together with the relation of divisibility, constitutes one of the basic examples of partially ordered sets, it is of considerable combinatorial importance.

Partitions of Sets of Integers

Another important area of combinatorial number theory is the study of what happens when the positive integers, or some finite subset thereof, are distributed among a bunch of smaller sets, often phrased in terms of "coloring" the integers. It is of particular interest to define a "largeness" or "density" property for sets of positive integers, and to show that, when the positive integers are divided into finitely many subsets, some subset is "large." For example, a result of Schur states that, for every positive integer k, there is some n with the following property: when the integers from 1 through n are divided into k subsets, some subset contains three distinct numbers such that one is the sum of the other two. Another significant example is Van der Waerden's theorem that, if the set of all positive integers is divided into finitely many subsets, there is some subset which contains arbitrarily long arithmetic progressions. In addition to finding properties that one subset must have, one also studies relations between the subsets. This study is particularly useful when one restricts one's attention to partitions of the integers into particular types of subsets, such as arithmetic progressions. A variety of combinatorial and number theoretic techniques - construction, induction, contradiction, direct proofs - are used in this area.

Additive Problems

One of the largest areas of combinatorial number theory - and one of the broadest, as it connects not only with combinatorics but also analysis and algebra - is additive number theory: the study of what happens when sets of integers are added together. One of the most famous unsolved problems in number theory is an additive one: the Goldbach conjecture, which says that every even number greater than 4 is the sum of two odd primes. Many combinatorial and algebraic techniques prove extremely valuable here; most notable is the pigeonhole principle, but other methods of approaching problems include classifying sets of numbers to be added, the use of minimal and maximal elements, counting sets in multiple ways, and the representation of a sum of numbers as a difference of two other sums.

Partitions and Related Topics

The deepest and most serious area of research in combinatorial number theory is concerned with partitions of a positive integer. A partition is a representation of an integer as a sum of other positive integers, called the parts. Two partitions with the same parts in a different order are considered the same. For example, 4 has 5 partitions: 4, 3 + 1, 2 + 2, 2 + 1 + 1, and 1 + 1 + 1 + 1. Many colorful results are concerned with the number of partitions of a positive integer or the number of partitions satisfying a particular condition. A famous result of Ramanujan (who did extensive work on partitions) asserts that the number of partitions of a number of the form 5k + 4 is always divisible by 5. Also important is Euler's "pentagonal number theorem," which provides a recurrence relation for calculating the number $p(n)$ of partitions of n: it equals the sum of $(-1)^{m+1}(p(n-\frac{m(3m-1)}{2}+p(n-\frac{m(3+1)}{2})$ overall positive integers m, where we take $p(0)=1$ and $p(n)=0$. Partitions played a crucial role in the recent proof of the alternating sign matrix conjecture.

Unlike many other areas of combinatorial number theory, there are a few specific techniques which are of considerable importance in examining partitions. One is the use of Ferrers diagrams, a method of representing partitions graphically; also useful is the technique of bijection, where one finds a function that turns one type of partition into another type of partition and thereby shows that the number of partitions of the first type equals the number of partitions of the second type. By far the most significant tool, however, is the use of the generating function. A generating function is a power series (or "infinite polynomial") in which the coefficients represent a significant sequence of numbers. By doing algebraic operations on entire series, one can obtain information about the individual coefficients. Sequences where the nth coefficients represents information about partitions of n happen to be particularly amenable to this kind of manipulation. There are non-partition-related problems as well which can be solved with generating-function techniques.

LARGE SET

In combinatorial mathematics, a large set of positive integers

$$S = \{s_0, s_1, s_2, s_3, \ldots\}$$

is one such that the infinite sum of the reciprocals

$$\frac{1}{s_0} + \frac{1}{s_1} + \frac{1}{s_2} + \frac{1}{s_3} + \cdots$$

diverges. A small set is any subset of the positive integers that is not large; that is, one whose sum of reciprocals converges.

Large sets appear in the Müntz–Szász theorem and in the Erdős conjecture on arithmetic progressions.

Examples:

- Every finite subset of the positive integers is small.
- The set $\{1,2,3,4,5,\ldots\}$ of all positive integers is known to be a large set; this statement is equivalent to the divergence of the harmonic series. More generally, any arithmetic progression (i.e., a set of all integers of the form $an + b$ with $a \geq 1$, $b \geq 1$ and $n = 0, 1, 2, 3, \ldots$) is a large set.
- The set of square numbers is small. So is the set of cube numbers, the set of 4th powers, and so on. More generally, the set of positive integer values of any polynomial of degree 2 or larger forms a small set.
- The set {1, 2, 4, 8, ...} of powers of 2 is known to be a small set, and so is any geometric progression (i.e., a set of numbers of the form of the form ab^n with $a \geq 1$, $b \geq 2$ and $n = 0, 1, 2, 3, \ldots$).
- The set of prime numbers has been proven to be large. The set of twin primes has been proven to be small.
- The set of prime powers which are not prime (i.e., all numbers of the form p^n with $n \geq 2$ and p prime) is a small set although the primes are a large set. This property is frequently used in analytic number theory. More generally, the set of perfect powers is small.
- The set of numbers whose expansions in a given base exclude a given digit is small. For example, the set $\{1,\ldots,6,8,\ldots,16,18,\ldots,66,68,69,80,\ldots\}$ of integers whose decimal expansion does not include the digit 7 is small. Such series are called Kempner series.
- Any set whose upper asymptotic density is nonzero, is large.

Properties

- Every subset of a small set is small.

- The union of finitely many small sets is small, because the sum of two convergent series is a convergent series. (In set theoretic terminology, the small sets form an ideal).
- The complement of every small set is large.
- The Müntz–Szász theorem states that a set $S = \{s_1, s_2, s_3, \ldots\}$ is large if and only if the set of polynomials spanned by $\{1, x^{s_1}, x^{s_2}, x^{s_3}, \ldots\}$ is dense in the uniform norm topology of continuous functions on a closed interval. This is a generalization of the Stone–Weierstrass theorem.

COVERING SYSTEM

In mathematics, a covering system (also called a complete residue system) is a collection

$$\{a_1 (\text{mod } n_1), \ldots, a_k (\text{mod } n_k)\}$$

of finitely many residue classes $a_i (\text{mod } n_i) = \{a_i + n_i x : x \in \mathbb{Z}\}$ whose union contains every integer.

The notion of covering system was introduced by Paul Erdős in the early 1930s.

The following are examples of covering systems:

$$\{0(\text{mod } 3), 1(\text{mod } 3), 2(\text{mod } 3)\},$$

and,

$$\{1(\text{mod } 2), 2(\text{mod } 4), 4(\text{mod } 8), 0(\text{mod } 8)\},$$

and,

$$\{0(\text{mod } 2), 0(\text{mod } 3), 1(\text{mod } 4), 5(\text{mod } 6), 7(\text{mod } 12)\}.$$

A covering system is called disjoint (or exact) if no two members overlap.

A covering system is called distinct (or incongruent) if all the moduli n_i are different (and bigger than 1).

A covering system is called irredundant (or minimal) if all the residue classes are required to cover the integers.

The first two examples are disjoint.

The third example is distinct.

A system (i.e., an unordered multi-set)

$$\{a_1 (\text{mod } n_1), \ldots, a_k (\text{mod } n_k)\}$$

of finitely many residue classes is called an m-cover if it covers every integer at least times, and an *exact* m-cover if it covers each integer exactly m times. It is known that for each $m = 2,3,\ldots$ there are exact m-covers which cannot be written as a union of two covers. For example,

$$\{1(\text{mod } 2); 0(\text{mod } 3); 2(\text{mod } 6); 0,4,6,8(\text{mod } 10);$$

$$1,2,4,7,10,13(\text{mod } 15); 5,11,12,22,23,29(\text{mod } 30)\}$$

is an exact 2-cover which is not a union of two covers.

The first example above is an exact 1-cover (also called an *exact cover*). Another exact cover in common use is that of odd and even numbers, or

$$\{0(\text{mod } 2), 1(\text{mod } 2)\}.$$

This is just one case of the following fact: For every positive integer modulus m, there is an exact cover:

$$\{0(\text{mod } m), 1(\text{mod } m), \ldots, m-1(\text{mod } m)\}.$$

Mirsky–Newman Theorem

The Mirsky–Newman theorem, a special case of the Herzog–Schönheim conjecture, states that there is no disjoint distinct covering system. This result was conjectured in 1950 by Paul Erdős and proved soon thereafter by Leon Mirsky and Donald J. Newman. However, Mirsky and Newman never published their proof. The same proof was also found independently by Harold Davenport and Richard Rado.

Primefree Sequences

Covering systems can be used to find primefree sequences, sequences of integers satisfying the same recurrence relation as the Fibonacci numbers, such that consecutive numbers in the sequence are relatively prime but all numbers in the sequence are composite numbers. For instance, a sequence of this type found by Herbert Wilf has initial terms

$$a_1 = 20615674205555510, a_2 = 3794765361567513.$$

In this sequence, the positions at which the numbers in the sequence are divisible by a prime p form an arithmetic progression; for instance, the even numbers in the sequence are the numbers a_i where i is congruent to 1 mod 3. The progressions divisible by different primes form a covering system, showing that every number in the sequence is divisible by at least one prime.

Boundedness of the Smallest Modulus

Paul Erdős asked whether for any arbitrarily large N there exists an incongruent covering system the minimum of whose moduli is at least N. It is easy to construct examples where the minimum of the moduli in such a system is 2, or 3 (Erdős gave an example where the moduli are in the set of the divisors of 120; a suitable cover is 0(3), 0(4), 0(5), 1(6), 1(8), 2(10), 11(12), 1(15), 14(20), 5(24),

8(30), 6(40), 58(60), 26(120)) D. Swift gave an example where the minimum of the moduli is 4 (and the moduli are in the set of the divisors of 2880). S. L. G. Choi proved that it is possible to give an example for $N = 20$, and Pace P Nielsen demonstrates the existence of an example with $N = 40$, consisting of more than congruences.

Erdős's question was resolved in the negative by Bob Hough. Hough used the Lovász local lemma to show that there is some maximum $N<10^{16}$ which can be the minimum modulus on a covering system.

Systems of Odd Moduli

There is a famous unsolved conjecture from Erdős and Selfridge: an incongruent covering system (with the minimum modulus greater than 1) whose moduli are odd, does not exist. It is known that if such a system exists with square-free moduli, the overall modulus must have at least 22 prime factors.

RESTRICTED SUMSET

In additive number theory and combinatorics, a restricted sumset has the form

$$S = \{a_1 + \cdots + a_n : a_1 \in A_1, \ldots, a_n \in A_n \text{ and } P(a_1, \ldots, a_n) \neq 0\},$$

where $A_1, \ldots, A_n$ are finite nonempty subsets of a field F and $P(x_1, \ldots, x_n)$ is a polynomial over F.

when $P(x_1, \ldots, x_n) = 1$, S is the usual sumset $A_1 + \cdots + A_n$ which is denoted by nA if $A_1 = \cdots = A_n = A$; when

$$P(x_1, \ldots, x_n) = \prod_{1 \le i < j \le n} (x_j - x_i),$$

S is written as $A_1 \dotplus \cdots \dotplus A_n$ which is denoted by $n^\wedge A$ if $A_1 = \cdots = A_n = A$. Note that $|S| > 0$ if and only if there exist $a_1 \in A_1, \ldots, a_n \in A_n$ with $P(a_1, \ldots, a_n) \neq 0$.

Cauchy–Davenport Theorem

The Cauchy–Davenport theorem named after Augustin Louis Cauchy and Harold Davenport asserts that for any prime p and nonempty subsets A and B of the prime order cyclic group Z/pZ we have the inequality

$$|A + B| \ge \min\{p, |A| + |B| - 1\}.$$

We may use this to deduce the Erdős–Ginzburg–Ziv theorem: given any sequence of $2n-1$ elements in Z/n, there are n elements that sums to zero modulo n. (Here n does not need to be prime.)

A direct consequence of the Cauchy-Davenport theorem is: Given any set S of $p-1$ or more nonzero elements, not necessarily distinct, of Z/pZ, every element of Z/pZ can be written as the sum of the elements of some subset (possibly empty) of S.

Kneser's theorem generalises this to general abelian groups.

Erdős–Heilbronn Conjecture

The Erdős–Heilbronn conjecture posed by Paul Erdős and Hans Heilbronn in 1964 states that $|2^\wedge A| \geq \min\{p, 2|A|-3\}$ if p is a prime and A is a nonempty subset of the field Z/pZ. This was first confirmed by J. A. Dias da Silva and Y. O. Hamidoune in 1994 who showed that:

$$|n^\wedge A| \geq \min\{p(F), n|A|-n^2+1\},$$

where A is a finite nonempty subset of a field F, and $p(F)$ is a prime p if F is of characteristic p, and $p(F) = \infty$ if F is of characteristic 0. Various extensions of this result were given by Noga Alon, M. B. Nathanson and I. Ruzsa, Q. H. Hou and Zhi-Wei Sun, and G. Karolyi.

Combinatorial Nullstellensatz

A powerful tool in the study of lower bounds for cardinalities of various restricted sumsets is the following fundamental principle: the combinatorial Nullstellensatz. Let $f(x_1,\ldots,x_n)$ be a polynomial over a field F. Suppose that the coefficient of the monomial $x_1^{k_1}\cdots x_n^{k_n}$ in $f(x\ ,\ldots,x\ )$ is nonzero and $k_1+\cdots+k_n$ is the total degree of $f(x_1,\ldots,x_n)$. If $A_1,\ldots,A_n$ are finite subsets of F with $|A_i|>k_i$ for $i=1,\ldots,n,$, then there are $a_1\in A_1,\ldots,a_n\in A_n$ such that $f(a_1,\ldots,a_n)\neq 0$.

The method using the combinatorial Nullstellensatz is also called the polynomial method.

GILBREATH'S CONJECTURE

Gilbreath's conjecture is a conjecture in number theory regarding the sequences generated by applying the forward difference operator to consecutive prime numbers and leaving the results unsigned, and then repeating this process on consecutive terms in the resulting sequence, and so forth. The statement is named after mathematician Norman L. Gilbreath who, in 1958, presented it to the mathematical community after observing the pattern by chance while doing arithmetic on a napkin. In 1878, eighty years before Gilbreath's discovery, François Proth had, however, published the same observations along with an attempted proof, which was later shown to be false.

Gilbreath observed a pattern while playing with the ordered sequence of prime numbers,

2, 3, 5, 7, 11, 13, 17, 19, 23, 29, 31, ...

Computing the absolute value of the difference between term n+1 and term n in this sequence yields the sequence,

1, 2, 2, 4, 2, 4, 2, 4, 6, 2, ...

If the same calculation is done for the terms in this new sequence, and the sequence that is the outcome of this process, and again *ad infinitum* for each sequence that is the output of such a calculation, the following five sequences in this list are:

1, 0, 2, 2, 2, 2, 2, 2, 4, ...

1, 2, 0, 0, 0, 0, 0, 2, ...

1, 2, 0, 0, 0, 0, 2, ...

1, 2, 0, 0, 0, 2, ...

1, 2, 0, 0, 2, ...

What Gilbreath—and François Proth before him—noticed is that the first term in each series of differences appears to be 1.

The Conjecture

Stating Gilbreath's observation formally is significantly easier to do after devising a notation for the sequences in the previous section. Toward this end, let $\{p_n\}$ denote the ordered sequence of prime numbers p_n, and define each term in the sequence $\{d_n^1\}$ by,

$$d_n^1 = p_{n+1} - p_n,$$

where n is positive. Also, for each integer k greater than 1, let the terms in $\{d_n^k\}$ be given by,

$$d_n^k = | d_{n+1}^{k-1} - d_n^{k-1} |.$$

Gilbreath's conjecture states that every term in the sequence $a_k = d_1^k$ for positive k is 1.

Verification and Attempted Proofs

As of 2013, no valid proof of the conjecture has been published. François Proth released what he believed to be a proof of the statement that was later shown to be flawed. Andrew Odlyzko verified that d_1^k is 1 for $k \leq n = 3.4 \times 10^{11}$, but the conjecture remains an open problem. Instead of evaluating n rows, Odlyzko evaluated 635 rows and established that the 635th row started with a 1 and continued with only 0s and 2s for the next n numbers. This implies that the next n rows begin with a 1.

In 1980, Martin Gardner published a conjecture by Hallard Croft that stated that the property of Gilbreath's conjecture (having a 1 in the first term of each difference sequence) should hold more generally for every sequence that begins with 2, subsequently contains only odd numbers, and has a sufficiently low bound on the gaps between consecutive elements in the sequence. This conjecture has also been repeated. However, it is false: for every initial subsequence of 2 and odd numbers, and every non-constant growth rate, there is a continuation of the subsequence by odd numbers whose gaps obey the growth rate but whose difference sequences fail to begin with 1 infinitely often. Odlyzko is more careful, writing of certain heuristic reasons for believing Gilbreath's conjecture that "the arguments above apply to many other sequences in which the first element is a 1, the others even, and where the gaps between consecutive elements are not too large and are sufficiently random." However, he does not give a formal definition of what "sufficiently random" means.

SZEMERÉDI'S THEOREM

In arithmetic combinatorics, Szemerédi's theorem is a result concerning arithmetic progressions in subsets of the integers. In 1936, Erdős and Turán conjectured that every set of integers A with positive natural density contains a k-term arithmetic progression for every k. Endre Szemerédi proved the conjecture in 1975.

Statement

A subset A of the natural numbers is said to have positive upper density if

$$\limsup_{n\to\infty} \frac{|A \cap \{1,2,3,\ldots,n\}|}{n} > 0.$$

Szemerédi's theorem asserts that a subset of the natural numbers with positive upper density contains infinitely many arithmetic progressions of length k for all positive integers k.

An often-used equivalent finitary version of the theorem states that for every positive integer k and real number $\delta \in (0,1]$, there exists a positive integer

$$N = N(k,\delta)$$

such that every subset of $\{1, 2, ..., N\}$ of size at least δN contains an arithmetic progression of length k.

Another formulation uses the function $r_k(N)$, the size of the largest subset of $\{1, 2, ..., N\}$ without an arithmetic progression of length k. Szemerédi's theorem is equivalent to the asymptotic bound

$$r_k(N) = o(N).$$

That is, $r_k(N)$ grows less than linearly with N.

Quantitative Bounds

It is an open problem to determine the exact growth rate of $r_k(N)$. The best known general bounds are

$$CN \exp\left(-n2^{(n-1)/2} \sqrt[n]{\log N} + \frac{1}{2n} \log\log N\right) \le r_k(N) \le \frac{N}{(\log\log N)^{2^{-2^{k+9}}}},$$

where $n = \lceil \log k \rceil$. The lower bound is due to O'Bryant building on the work of Behrend, Rankin, and Elkin. The upper bound is due to Gowers.

For small k, there are tighter bounds than the general case. When $k = 3$, Bourgain, Heath-Brown,

Szemerédi, and Sanders provided increasingly smaller upper bounds. The current best bounds are

$$N2^{-\sqrt{8\log N}} \le r_3(N) \le C\frac{(\log\log N)^4}{\log N}N$$

due to O'Bryant and Bloom respectively.

For $k = 4$, Green and Tao proved that

$$r_4(N) \le C\frac{N}{(\log N)^c}$$

for some $c > 0$.

Extensions and Generalizations

A multidimensional generalization of Szemerédi's theorem was first proven by Hillel Furstenberg and Yitzhak Katznelson using ergodic theory. Timothy Gowers, Vojtěch Rödl and Jozef Skokan with Brendan Nagle, Rödl, and Mathias Schacht, and Terence Tao provided combinatorial proofs.

Alexander Leibman and Vitaly Bergelson generalized Szemerédi's to polynomial progressions: If $A \subset \mathbb{N}$ is a set with positive upper density and $p_1(n), p_2(n), \dots, p_k(n)$ are integer-valued polynomials such that $p_i(0) = 0$, then there are infinitely many $u, n \in \mathbb{Z}$ such that $u + p_i(n) \in A$ for all $1 \le i \le k$. Leibman and Bergelson's result also holds in a multidimensional setting.

The finitary version of Szemerédi's theorem can be generalized to finite additive groups including vector spaces over finite fields. The finite field analog can be used as a model for understanding the theorem in the natural numbers. The problem of obtaining bounds in the k=3 case of Szemerédi's theorem in the vector space $\mathbb{F}_3^n$ is known as the cap set problem.

The Green–Tao theorem asserts the prime numbers contain arbitrary long arithmetic progressions. It is not implied by Szemerédi's theorem because the primes have density 0 in the natural numbers. As part of their proof, Ben Green and Tao introduced a "relative" Szemerédi theorem which applies to subsets of the integers (even those with 0 density) satisfying certain pseudorandomness conditions. A more general relative Szemerédi theorem has since been given by David Conlon, Jacob Fox, and Yufei Zhao.

The Erdős conjecture on arithmetic progressions would imply both Szemerédi's theorem and the Green–Tao theorem.

ERDŐS–GRAHAM PROBLEM

In combinatorial number theory, the Erdős–Graham problem is the problem of proving that, if the set {2, 3, 4, ...} of integers greater than one is partitioned into finitely many subsets, then one of the subsets can be used to form an Egyptian fraction representation of unity. That is, for every $r > 0$, and every r-coloring of the integers greater than one, there is a finite monochromatic subset S of these integers such that $\sum_{n\in S}\frac{1}{n}=1$.

Paul Erdős and Ronald Graham conjectured that, for sufficiently large r, the largest member of S could be bounded by b^r for some constant b independent of r. It was known that, for this to be true, b must be at least e.

Ernie Croot proved the conjecture. The value Croot gives for b is very large: it is at most e^{167000}. Croot's result follows as a corollary of a more general theorem stating the existence of Egyptian fraction representations of unity for sets C of smooth numbers in intervals of the form $[X, X^{1+\delta}]$, where C contains sufficiently many numbers so that the sum of their reciprocals is at least six. The Erdős–Graham conjecture follows from this result by showing that one can find an interval of this form in which the sum of the reciprocals of all smooth numbers is at least $6r$; therefore, if the integers are r-colored there must be a monochromatic subset C satisfying the conditions of Croot's theorem.

ZERO-SUM PROBLEM

In number theory, zero-sum problems are certain kinds of combinatorial problems about the structure of a finite abelian group. Concretely, given a finite abelian group G and a positive integer n, one asks for the smallest value of k such that every sequence of elements of G of size k contains n terms that sum to 0.

The classic result in this area is the 1961 theorem of Paul Erdős, Abraham Ginzburg, and Abraham Ziv. They proved that for the group $\mathbb{Z}/n\mathbb{Z}$ of integers modulo n, $k = 2n-1$.

Explicitly this says that any multiset of $2n - 1$ integers has a subset of size n the sum of whose elements is a multiple of n, but that the same is not true of multisets of size $2n - 2$. (Indeed, the lower bound is easy to see: the multiset containing $n - 1$ copy of 0 and $n - 1$ copies of 1 contains no n-subset summing to a multiple of n.) This result is known as the Erdős–Ginzburg–Ziv theorem after its discoverers. It may also be deduced from the Cauchy–Davenport theorem.

More general results than this theorem exist, such as Olson's theorem, Kemnitz's conjecture and the weighted EGZ theorem.

ZNÁM'S PROBLEM

Znám's problem is a problem posed by the Slovak mathematician Stefan Znám in 1972 asking whether, for all integers $k \geq 2$, there exist k integers $x_1,\ldots,x_k$ all greater than 1 such that x_i is a proper divisor of $x_1\ldots x_k / x_i + 1$ for each $i = 1,\ldots,k$. The answer is negative for $2 \leq k \leq 4$ and affirmative for $k \geq 5$. Sun Qi also gave a lower bound for the number $Z(k)$ of solutions.

All solutions for $5 \le k \le 8$ have now been computed, summarized in the table below. The numbers of solutions for $n = 2$, 3, ... terms are 0, 0, 0, 2, 5, 15, 93, ... and the solutions themselves.

Cao and Sun showed that $Z(11) \ge 5$ and Cao and Jing that there are ≥ 39 solutions for $n \ge 12$. A solution for $k = 13$ was found by Girgensohn in 1996: 3, 4, 5, 7, 29, 41, 67, 89701, 230865947737, 5726348063558735709083, followed by large numbers having 45, 87, and 172 digits.

It has been observed that all known solutions to Znám's problem provide a decomposition of 1 as an Egyptian fraction

$$\frac{1}{x_1} + \frac{1}{x_2} + \ldots + \frac{1}{x_k} + \frac{1}{x_1 \cdots x_k} = 1.$$

Conversely, every solution to this Diophantine equation is a solution to Znám's problem, unless $x_i = x_1 \ldots x_k / x_i + 1$ for some i.

References

- Guo, Song; Sun, Zhi-Wei (2005). "On odd covering systems with distinct moduli". Adv. Appl. Math. 35 (2): 182–187. Arxiv:math/0412217. Doi:10.1016/j.aam.2005.01.004. MR 2152886
- Soifer, Alexander (2009). The Mathematical Coloring Book: Mathematics of Coloring and the Colorful Life of its Creators. With forewords by Branko Grünbaum, Peter D. Johnson, Jr. And Cecil Rousseau. New York: Springer. Pp. 1–9. Doi:10.1007/978-0-387-74642-5. ISBN 978-0-387-74640-1. MR 2458293
- Hou, Qing-Hu; Sun, Zhi-Wei (2002). "Restricted sums in a field". Acta Arithmetica. 102 (3): 239–249. Bibcode:2002acari.102..239H. Doi:10.4064/aa102-3-3. MR 1884717
- Guy, Richard K. (2004). Unsolved Problems in Number Theory. Problem Books in Mathematics (3rd ed.). Springer-Verlag. P. 42. ISBN 0-387-20860-7. Zbl 1058.11001
- Croot, Ernest S., III (2003). "On a coloring conjecture about unit fractions". Annals of Mathematics. 157 (2): 545–556. Arxiv:math.NT/0311421. Doi:10.4007/annals.2003.157.545. MR 1973054
- Znamsproblem: wolfram.com, Retrieved 13 May, 2019

PERMISSIONS

All chapters in this book are published with permission under the Creative Commons Attribution Share Alike License or equivalent. Every chapter published in this book has been scrutinized by our experts. Their significance has been extensively debated. The topics covered herein carry significant information for a comprehensive understanding. They may even be implemented as practical applications or may be referred to as a beginning point for further studies.

We would like to thank the editorial team for lending their expertise to make the book truly unique. They have played a crucial role in the development of this book. Without their invaluable contributions this book wouldn't have been possible. They have made vital efforts to compile up to date information on the varied aspects of this subject to make this book a valuable addition to the collection of many professionals and students.

This book was conceptualized with the vision of imparting up-to-date and integrated information in this field. To ensure the same, a matchless editorial board was set up. Every individual on the board went through rigorous rounds of assessment to prove their worth. After which they invested a large part of their time researching and compiling the most relevant data for our readers.

The editorial board has been involved in producing this book since its inception. They have spent rigorous hours researching and exploring the diverse topics which have resulted in the successful publishing of this book. They have passed on their knowledge of decades through this book. To expedite this challenging task, the publisher supported the team at every step. A small team of assistant editors was also appointed to further simplify the editing procedure and attain best results for the readers.

Apart from the editorial board, the designing team has also invested a significant amount of their time in understanding the subject and creating the most relevant covers. They scrutinized every image to scout for the most suitable representation of the subject and create an appropriate cover for the book.

The publishing team has been an ardent support to the editorial, designing and production team. Their endless efforts to recruit the best for this project, has resulted in the accomplishment of this book. They are a veteran in the field of academics and their pool of knowledge is as vast as their experience in printing. Their expertise and guidance has proved useful at every step. Their uncompromising quality standards have made this book an exceptional effort. Their encouragement from time to time has been an inspiration for everyone.

The publisher and the editorial board hope that this book will prove to be a valuable piece of knowledge for students, practitioners and scholars across the globe.

INDEX